Unveiling the Power of Women in Irrigation

Mary Okumu Jimoh · Sagie Narsiah

Unveiling the Power of Women in Irrigation

Challenges, Resilience, and the Ecofeminist Connection

palgrave
macmillan

Mary Okumu Jimoh
Sociology
University of KwaZulu-Natal
Durban, South Africa

Sagie Narsiah
Geography
University of KwaZulu-Natal
Durban, South Africa

ISBN 978-3-031-94173-3 ISBN 978-3-031-94174-0 (eBook)
https://doi.org/10.1007/978-3-031-94174-0

Cover illustration: Marina Lohrbach_shutterstock.com

This Palgrave Macmillan imprint is published by the registered company Springer Nature Switzerland AG
The registered company address is: Gewerbestrasse 11, 6330 Cham, Switzerland

The late James Oyiengo Okumu and Lenny and Saras Narsiah

Preface

This book originated from the need to share knowledge from my Phd thesis. My thesis delved into the contribution of women into irrigated farming systems in Kenya. There are many accounts of women in the agricultural sector. However, their role within the sector although acknowledged is not vocalized in a manner that showcases their daily struggles. The discussion of women farmers in the Global South is pronounced in the literature but not loud enough for change to be realized. During my research, I encountered many authors acknowledging the struggles of women farmers and the lack of empirical data to support their observations. The story has not changed much since I graduated. Although, some progress has been made over the years, more still needs to be done to encourage and support women in the agricultural sector.

There are six key themes with gender as the key element within each theme: women and agriculture, women and land tenure systems, water and irrigation, financial resources and assets, education and technology and agricultural politics. The issues affecting women are diverse. Specifically issues on land ownership and the accessibility to natural resources such as water. Interestingly, gaining access to these resources is dependent on the most crucial natural resource of all land. In addition, access to these resources is also important when one seeks to secure financial accessibility of assets and loans from financial institutions. The absence of basic education and technology hinders and complicates increased agricultural production for women due to the many roles they take up in the home such as unpaid care work. Lastly, politics, or agricultural politics is in every facet of society, be it in the general society or within a rural village. Politics plays its part in having one gain access, to or manage any natural resource, furthermore, differences emerge when politics does not seem to favour a group in society or the community at large.

Why is gender central to changing the agricultural landscape in Africa? Women are involved in every societal aspect of the community, from being house managers, mothers, wives, sisters and aunts, they are linked to every aspect of life. Ecofeminism is a theory that was developed in 1974 by the French feminist, Françoise d'Eaubonne, in her path-breaking book, '*Le Féminismeou la Mort*.' In her book she advocates the important link that women have with nature, and that women are held back due to the patriarchal nature that is present within society. It is because of this that women fight to protect nature.

Notable feminists around the world include the late Prof. Wangari Maathai from Kenya, the first African woman to win a Nobel Peace Prize for her astounding work in forest conservation in Kenya. The late Prof. Maathai channelled her work through the Green Belt Movement in Kenya. She was able to build a network of women based in rural Kenya, who fought the dwindling forest numbers in Kenya in the late 1990s, through national and political activism. She mobilized the women to plant trees and indigenous plants that were dwindling in number and fight deforestation that was destroying the water catchment areas within the Kenyan Highlands. Other notable ecofeminists include Vandana Shiva from India, who has called for the fight against capitalism that seeks to exploit the poor especially women.

This book, therefore seeks to draw on the empirical data gathered on the struggles that women face in irrigated agriculture in Kenya, but also draw from other examples within Africa, with a view to highlight the different experiences and challenges within the sector. Secondly, the need to highlight the impact of climate change on the agricultural sector and its effects on farmers, especially women farmers, who have to contend with migrations, due to famine and drought that affects both their crops and livelihoods. Thirdly, the importance of including gender as a key aspect when it comes to the management of natural resources, and lastly the need for agricultural policies, specifically targeted on natural resources such as land and water. The reasoning behind this is that women are still viewed as second-class citizens when it comes to agriculture and propelling them as individuals who have a strong impact in the sector, can be the beginning of change the agricultural sector in Africa.

In Chapter 1, we provide the rationale for the role of women in agriculture, while highlighting the different aspects of natural resource management that affect women in irrigated agriculture. We then follow up with an introduction to ecofeminism, with an overview of the important link women have with nature. We also discuss the connection women have with nature. This expands on the role of women in irrigated agriculture, discussing the inequalities women face and the changes that have taken place within the development space in fighting the disparities women face in agriculture. This is assayed into the core issues that affect women, discussing land, water, financial resources, technology and education, markets and labour. We highlight the importance of bridging the gender divide in the sector and conclude the chapter with a brief account of women's key contributions to the agricultural sector.

In Chapter 2 we discuss the role of the commons in managing natural resources and the impact this has on women. The commons is an economic theory popularized by the late Nobel laureate Elinor Ostrom, which seeks to highlight the pros and cons of natural resource management within different economic landscapes. In this chapter, we specifically focus on women and draw upon their experiences in managing natural resources and the impact this has on the resources and them. In addition, there is a follow-up discussion on financial resources, via capitalist practices that disadvantage women and natural resources. The discussion reflects on the examples of ecofeminists like Vandana Shiva, Mies and the late Nobel Laureate Prof. Wangari Maathai to highlight the challenges women face when managing these resources. The chapter concludes with a summary of the importance of managing natural resources.

In Chapter 3, we examine the impact of climate change on both women and men, highlighting the challenges it poses for farming and the strategies they can employ to mitigate its effects. We explore these impacts by discussing the specific challenges women face regarding water accessibility and emphasizing the importance of adopting both traditional and modern methods to enhance the resilience and sustainability of farming communities. The chapter further highlights the impact of climate change on women's financial accessibility and underscores the role of education in helping them adopt climate-smart farming practices. Additionally, it emphasizes the need to challenge and transform socio-cultural norms, which is crucial in building communities that are more resilient to the effects of climate change. The chapter concludes by stressing the importance of educating and training more women in climate change mitigation while encouraging them to leverage their deep connections with nature to empower themselves both economically and socially.

Chapter 4 explores the critical role of advancing women's participation in the agricultural sector and its impact on increasing food security. It emphasizes the need to integrate both women and youth into agricultural production, recognizing that their involvement is essential for strengthening food security—not just at the community level but also on a national scale. The chapter also examines the challenges women face across various aspects of agriculture, including economic empowerment, investment barriers, shifting farm structures and water security. It underscores the importance of policy reforms and social norm changes in addressing these challenges and creating a more significant impact on food security

Chapter 5 expands the discussion on gender and agricultural policies, examining their role in closing the gender gap in agriculture across Africa. The chapter specifically focuses on the impact of gender-responsive agricultural policies in Kenya and the challenges associated with their implementation. Furthermore, we explore potential interventions for developing inclusive agricultural policies that effectively address gender disparities. Recognizing the critical role of politics in policy formulation and implementation, we also analyze the influence of agricultural politics on decision-making processes. Lastly, we highlight the legacy of the late Nobel Laureate Prof. Wangari Maathai, particularly her political struggles to safeguard green spaces in Kenya. Her advocacy through the Green Belt Movement has left a lasting impact on the ecofeminist movement, reinforcing the intersection of environmental conservation, gender equity and political activism.

Chapter 6 provides a critical analysis of women in irrigation, their resilience, and the ecofeminist connection, with a focus on the Ahero Irrigation Scheme in Kenya as a case study. To establish a contextual foundation, we first trace the history of irrigation in Kenya, offering insight into the development of the Ahero Irrigation Scheme, the types of crops cultivated, and the challenges faced by farmers. The chapter is structured around the six key themes introduced in 1, presenting empirical findings aligned with these themes. Additionally, we examine the research methodology used for data collection and conclude with an in-depth analysis of the study's key findings.

In Chapter 7, we delve into the future of agriculture, discussing the impact of the fourth revolution, (the 4.0. revolution). We broaden discussions on the fourth revolution by highlighting the challenges women may face about land, finances and education and conclude with a rounding up of the impact of the fourth revolution in Africa.

Durban, South Africa
KwaZulu-Natal, South Africa

Mary Okumu Jimoh
Sagie Narsiah

Acknowledgements I would like to express my deepest gratitude to Prof. Sagie for his unwavering support and encouragement throughout the journey of writing this book. His guidance pushed me beyond my limits, and for that, I am truly grateful. A heartfelt thank you to Palgrave for their support, especially Rachel Ballard and Abarna Antonyraj, whose assistance was invaluable. To my family my late father, James Okumu, whose love and belief in all his children continue to inspire me, and my mother, Rachel, who never stopped believing in me, even when I doubted myself. To my sisters, Serah, Olivia, Nicole and Natalie, for their endless stories that filled my days with laughter, and my brothers, Hamisi and Martin, for their constant support. To my beloved in-laws, who embraced me from the very first day Taliat, Olubisi, Lanre, Tosin, Dr. Tosin, Tosin 2 and the Olushis-Wale, Lolade, Shade and Ebenezer your love and encouragement mean the world to me. To my dear friends, Samy, Sumeya, Sahara, Somawina, Ajo, Rao and Mwajuma, thank you for your unwavering friendship. A special appreciation to my mentor, Cleopatra Agoro, for her invaluable guidance. My immense gratitude to Dr. Joan Jaganyi for guiding me from the beginning. To my son, Fabian Olawale Oyiengo, who watched me write this book you are my greatest inspiration. And finally, to my best friend and love, Temitope Jimoh your support and belief in me made this journey possible.

Sagie Narsiah would like to acknowledge all those who contributed to the fruition of this project. Special thanks to Palgrave for accepting the proposed project and supporting academics from our 'part of the world.' During the project Mary lost her dear Dad and I lost my beloved Mum. We acknowledge their supreme sacrifice in our collective journey through life. It is with this in mind that we dedicate this book to their memory. To my special ones Sarah, Janessa and Catalena it is a joy to see you flourish. And, to Maya thanks for sharing my life always for love. To my sisters Caroline and Leanne, thank you for your support.

Competing Interests The authors have no competing interests to declare that are relevant to the content of this manuscript.

Contents

Acronyms

ACBF	African Capacity Building Foundation
AECM	Agricultural Extension and Communication Management
AFDB	African Development Bank
AGRA	Alliance for Green Revolution in Africa
AWARD	African Women in Agricultural Research and Development
AWM	Agricultural Water Management
CAAP	Comprehensive African Agricultural Development
CDC	Cameroon Development Corporation
CEDAW	Convention on the Elimination of all forms of Discrimination Against Women
CGIAR	The Consultative Group on International Agricultural Research
CLA	Community Land Act
CON	Constitution of Kenya Law
COP	Conference of the Parties
COVID-19	Corona Virus
CSA	Climate Smart Agriculture
DAWN	Development Alternative with Women for a New Era
EIAR	Ethiopian Institute of Agricultural Research
ENABLE	Empowering Novel Agribusiness Led Employment
FAO	Food and Agricultural Organisation
FISP	Farm Input Subsidy Program
GAD	Gender and Development
GBM	Green Belt Movement
GED	Gender Equality and Development
GEWE	Gender Equality and Women's Empowerment
GPS	Global Positioning System

GRASP	Gender Responsive Systems Policy Fellowship
GTA	Gender Transformative Approach
HIV	Human Immunodeficiency Virus
IFAD	International Fund for Agricultural Development
IPCC	Intergovernmental Panel on Climate Change
IWMI	International Water Management Institute
KCC	Kenya Co-operative Creameries
KMC	Kenya Meat Commission
LA	Land Act
LAPO	Lift Above Poverty Organization
LGP	Lorong Garden Program
LLL	Laser land leveling
LSMS	Living Standards Measurement Study
NARC	National Rainbow Coalition
NCWK	National Council of Women in Kenya
NFDP	National Fadama Development Project
NGO	Non-Governmental Organization
NIB	National Irrigation Board
SAKSS	Strategic Analysis Knowledge Support System
SDG	Sustainable Development Goals
SI	Sustainable Intensification
SMTP	Second Medium Term Plan
SSA	Sub-Saharan Africa
STEM	Science Technology Engineering and Mathematics
UK	United Kingdom
UN	United Nations
USAID	United States Agency for International Development
USD	United Stated Dollars
VWC	Voice of Women's Contribution
WAD	Women in Development
WEAI	Women's Empowerment in Agricultural Index
WEF	World Economic Forum
WELF	Water, Energy, Land and Food
WID	Women in Development
WLF	Water, Land, Food Security
YIYA	Youth Inspiring Youth in Agriculture

List of Figures

CHAPTER 1

Women's Role in Agriculture

1.1 Introduction

The role of women in agriculture globally is important, not just for academics, but for those working in the agricultural sector. Agriculture is one of the economic activities that can uplift poor people in the Global South. However, to date, investment in agriculture is low, and those practising within the industry still need help to receive inputs and other resources that are key in helping them grow the sector. Additionally, agriculture as a sector is facing numerous challenges that are arising as a result of climate change. The effects of climate change are visible globally. The crisis is escalating affecting many food systems worldwide. It is therefore important that those who work within the sector in any capacity, be it as farmers, Governments, researchers, suppliers or consumers take interest in improving the agricultural sector for a more sustainable system that can feed not only the world but local communities.

This book seeks to discuss the different aspects within the agricultural sector particularly the impact of gender. Specifically, it will highlight women's key role within the sector and their challenges. The book will use an ecofeminist lens, showcasing the link women have with nature, its importance, and how this link can make agriculture more sustainable for future generations.

Twenty percent of the total land cultivated worldwide receives irrigation water to produce about 40 percent of the world's food. It is

M. O. Jimoh and S. Narsiah, *Unveiling the Power of Women in Irrigation*, https://doi.org/10.1007/978-3-031-94174-0_1

interesting to note that sub-Saharan Africa has just 4 percent of this figure, the lowest percentage of irrigated land to the total area cultivated globally (Nakawuka et al., 2017). In addition, it has the highest depth of food deficit that is the highest amount of energy needed for people who are undernourished to attain the average dietary requirements. In the year 2000, 41 percent of the cultivated area in Asia was under irrigation an area tenfold the irrigated area in sub-Saharan Africa. Irrigated agriculture with sustainable farming practices in sub-Saharan Africa can create avenues to replicate circumstances that have led to increased yields as seen in the Asian continent in the last five decades. It is important to note that irrigation, mechanized agriculture, the use of enhanced seed varieties and farm supplies in the 1960s such as fertilizers and pesticides significantly contributed to the Green Revolution in Asia (Nakawuka et al., 2017).

Irrigation cannot be a basic survival practice as it requires heavy investment due to the construction of water holding or diversion facilities that are connected to water channels. Irrigation schemes need to have adequate personnel and institutional procedures that can ensure equitable distribution of water by carrying out repairs and proper payment of water resources. Expansion of irrigation systems shifts some farmers and draws new farmers leading to societal and political conflicts over land rights (Van Rooyen, 2017). The use of new innovations can enhance water efficiency in irrigation farms if development programs are encouraged. However, they must operate within complex institutional landscapes. The agricultural sector insub–Saharan Africa is undervalued with regard to critical structural settings and hence, misjudged with regard to the cost of innovation investment. Thus, exerting influence over water comes at a great price, both in structural and economic terms, and hence, the reason why many irrigation schemes are seen as human development-related failures. Irrigation systems, need to be seen as complex systems, that have different system levels and subsystems. This is because various stakeholders need to work together to share information and do so in a structured way that allows for functionality (Van Rooyen, 2017).

In Kenya, since the 1970s the irrigation sector has grown with the help of privately-owned large farms as well as small-scale farmers. Most of these farmers have received increased support from government and non-governmental organizations (NGOs) over the years which has led to increased agricultural production of horticultural and export crops via surface-fed irrigation systems (Nakawuka et al., 2017). Twenty kilometres, east of Kisumu city, on the eastern margin of the Winam Gulf of

Lake Victoria, sits the Ahero Irrigation scheme that is managed by the Kenya National Irrigation Board Western Regional Project.

The scheme has a total of 533 farming households, within the 878 hectares which is divided into 12 blocks ranging from 31 to 115 ha. Farming households have each been allocated 1.6 ha of paddy field. The scheme is supplied with water for irrigation from River Nyando, where rice is planted in two seasons annually which often coincide with the local rainfall patterns; one crop is harvested in July and the other in January (Makini et al., 2017).

The low productivity of public irrigation schemes in Kenya and the inefficient use of water is a major concern. Rice farming in Ahero, West Kano, Bunyala and Mwea is very low, thus, making rice production very low during drought periods and rainy seasons when continuous flooding methods utilize a lot of water. Lefore et al. (2019b) assert that politically connected individuals who are well-resourced will reduce the growth of irrigation. Current economic environments prevent irrigation from reaching the number of people who can benefit both directly and indirectly from equitable access to irrigation. This is because poor farmers are disadvantaged and their control over natural resources is more entrenched. It is therefore important to ensure equitable access to irrigation technologies, practices and complementary inputs that are sustainable to intensify production (Muema, 2018). Below are some of the challenges that need to be addressed to improve and guarantee the sustainable expansion of irrigation farming amongst women farmers.

1.2 Land Tenure Systems

The potential benefits of increased female control over land is often undermined by prevailing land rights systems that tend to favour men in sub-Saharan Africa. The access women have to farming inputs and land is restricted despite its critical importance. Despite, this being a critical aspect (Djurfeldt et al., 2018a). Women need to have rights-based advantages that ensure their land is secure. For example, tenure security is preserved in the event of divorce, allowing women to maintain their agricultural livelihoods strengthening their autonomy and voice both within and beyond the household. Moreover, enhancing women's control over land brings economic benefits. It increases their chances of accessing credit and capital-intensive technologies, as land can be used as collateral. Access to productive assets like land is essential in breaking the cycle of chronic poverty. However, critical perspectives question the feasibility of individualizing women's land rights within systems dominated by customary tenure. This skepticism stems from the fact that indigenous customary law often holds greater authority than statutory national law, as it is deeply rooted in longstanding social and cultural norms (Matheson & Ashleigh, 2019). As Nakawuka et al. (2017: 7) emphasize, 'Access to land and water resources is linked to poverty, as the world's poorest are also reported to have the least access to land and water.'

1.3 Water and Irrigation

Harrison and Mdee (2017: 408) describe the history of irrigation in sub-Saharan Africa as 'troubled, with an uncertain future.' They attribute this to the continent's repeated failure to learn from past mistakes. Despite some progress, the overall impact has been limited due to a recurring cycle of 'build-neglect-rebuild.' Additionally, irrigation efforts often go unrecognized by those promoting and assessing agricultural development, as they frequently fall outside formal projects and schemes. However, in recent years, there has been renewed interest in irrigation, particularly through efforts to rehabilitate past failures. This resurgence is largely driven by small-scale, community-based farmers who are actively advancing irrigation initiatives. In addition, it is important that gender and water are recognized as interdependent, so as to advance the international development agenda.

Integrating gender considerations into water resource management in developing countries has proven challenging due to its broad and complex scope. Factors such as the cost of water, its impact on economic performance, and the damage caused by water-related disasters create financial uncertainties, making it difficult to secure funding and further gender mainstreaming efforts. The primary objective of water resource management is to ensure water security, which refers to the availability of sufficient and safe water for health, livelihoods, ecosystems, and production while minimizing water-related risks to people, environments, and economies. Water security plays a critical role in social and economic development, encompassing effective governance, responsible stewardship, and the necessary infrastructure to harness water's benefits. Achieving water security requires strategic financial management to invest in infrastructure and development. This process must incorporate a gender perspective, ensuring that policies, planning, implementation, coordination, and evaluation consider the diverse ways in which men and women use water (Odonkor et al., 2018).

Women are often excluded when it comes to decision-making with regard to water. This is because they are yet to be included in the decision-making process in the community, regional and national levels. At the household level women are viewed as being custodians for water, with the same being applied with regard to water for agricultural use. It is for this reason that they need to be empowered to take an effective part in decision-making at all levels of water resource management (Odonkor et al., 2018).

1.4 Financial Resources and Assets

In Kenya, the financial sector has a wide array of avenues which can be used by individuals to access credit. However, this is not the case for farmers. The avenues of credit available require collateral, which is not easily available to farmers. In addition, the credit facilities available to farmers are for the purchase of seedlings and fertilizers and not for the purchase of agricultural technologies. As a result, many farmers rely on their personal savings to buy irrigation equipment. Agriculture is a risky venture owing to many issues, including pests, diseases and volatile food prices and thus without insurance, to cover one against such incidents, the return on investment may be none, hence this is why banks and lending institutions are reluctant to offer farmers loans (Nakawuka et al., 2017).

In Nigeria's Cross River State, for example, women do not have adequate access to loan facilities and if they do uptake is low. Organization's like Lift Above Poverty Organization (LAPO) or National Fadama Development Project (FADAMA) are unable to support women farmers with loans. Furthermore, the funds available do not exceed above one hundred thousand Naira N100, 000.00 which is the equivalent of ($300). These amounts are unable to facilitate high production farming. This has been attributed to the terms and conditions of the loan facilities, which can result in the borrower losing some of their assets. With such conditions, many women are skeptical of taking up loans, hence no investments are seen in food production (Agbor & Eteng, 2018).

'In addition, low customer numbers make it difficult to offer certain services to customers in rural areas, due to poor record keeping and inadequate customer follow-ups' (Nakawuka et al., 2017: 9). This is because commercial banks and microfinance institutions do not open branches in remote rural areas as business related costs and potential liabilities are high. Furthermore, when they are present few agricultural loans are offered compared to loans in other services like trade. Financial institutions founded on the basis of assisting women often do not assist, thus exacerbating the problems further as Agbor and Eteng (2018) claimed. The application process is long and tedious, and even when approved, the lending rates on the loans are high making repayments difficult especially for women. To make matters worse, the political elites, use their influence to determine who can access the loans. It is therefore difficult to receive the funds if one is not connected to the elites as is often the case, thus, creating a hinderance for women in the sector. These harsh rules are seen to be part of the loan application processes in the formal sector. Hence, forcing rural women farmers to seek loans from the informal sector (e.g., Osusu groups- finance groups). In Imo State Nigeria, a study by Agbo and Ihemezie (2015) found out that 74.5 percent of the farmers in the research gained loans from the informal sector, while 25.5 percent gained access to loans from the formal sector. The government is obligated to members of the public to ensure that they receive financial services easily. In addition, such loans should not be provided to individuals based on their political connections but rather on the farmers' abilities to make loan repayments based on their capacities. For Agbor and Eteng, 'this will go a long way in encouraging good credit practices that would eventually lead to the purchase of improved varieties and other resources needed to improve agricultural production' (Agbor & Eteng, 2018: 395).

1.5 Technology and Education

'Agricultural technology enhances sustainable production of food and is critical for sustainable food security and economic development of all nations' (Oluwatusin, 2017: 72). The development of new farming innovations in agriculture continues to have interest, that dates back to the twentieth century; with current innovations being highly relevant to small-scale farmers in the Global South. Thus, there is need to develop irrigation systems and structures, markets and loan facilities within localities where there is no development so as to improve production. Over the years, studies have been conducted on the ways in which new innovations are being taking up by small scale farmers and the impact of this on crop production and finances. This research has been focused in the Global South. Data has shown that the uptake of new innovations is poor and slow. This has been attributed to the poor grasp of innovations. There exists a significant opportunity to expand Africa's irrigation potential (Oluwatusin, 2017). This is important because 'irrigation technology has been viewed as the missing link between the current state of agriculture in sub-Saharan Africa and its transition to a developed state' (Macharia & Carpio, 2018: 4). The investment and development of rural infrastructure, water management and effective policies is seen as critical in changing the poverty cycle that affects small-scale farmers in Africa. Policymakers and governments need to understand the critical link between agriculture and water management to ensure Africa is food secure (Macharia & Carpio, 2018).

Agricultural mechanization stands for the application of mechanical technology aimed at enhancing productivity and surpassing human capacity. Most countries that have achieved this feat have done so through heavy investments in mechanized agriculture. Agricultural mechanization is key to enabling farmers to intensify their production, enhance their efficiency, and improve their productivity and quality of life (Mukasa et al., 2017). This is because the use of machinery enables the farmers to improve their health as they are not engaged in heavy agricultural activity and are less likely to fall sick, thus reducing their costs in health care. Furthermore, family expenditure with regard to education and food can increase, leading to better health, farm production and increased family income (Mukasa et al., 2017).

Transforming small-scale agriculture in sub-Saharan Africa requires a change from rainfed subsistence agriculture to more highly commercial productive agricultural systems. This is because regions that have taken up irrigation have the best poverty reduction impacts on their communities (Macharia & Carpio, 2018). Furthermore, crop yields in these areas have increased and there is a greater demand for agricultural labour. Furthermore, irrigation impacts different beneficiaries differently. The impact can vary depending on the technology used, the crop cultivated, and the layout of the distribution system (head enders (farmers near the water source) or tail-enders (farmers further from the water source). In addition, the institutional rules governing access to water and use must be well laid out to ensure that all beneficiaries work within the system provided (Macharia & Carpio, 2018).

'Despite the many efforts by organizations and African governments, the rate of adoption of modern technologies in Africa is still considerably low. This they blame on the low literacy levels that hinder the adoption of new technologies. The participation of young people and women in advocating for critical issues affecting Africa is important, because their participation can improve food security and improve the economy which in turn reduces poverty in the continent' (Mukasa et al., 2017: 7–11). Waiganjo and Waweru (2018: 25) argue 'that it is imperative to educate women with basic agricultural knowledge and skills. This is not only important for Kenya, but is key if the whole of sub-Saharan Africa, would like to achieve the realizations of the 2030 agendas, which seek to improve the quality of life to citizens. To have a skilled workforce in agriculture it is important that the subject is taught in schools so that the next generation can improve agricultural production. Waiganjo and Waweru (2018) argued that because agriculture is a science, it should be taught using teaching approaches that allow students to acquire not only the knowledge, but skills that can build the sector and promote better knowledge of agriculture. The challenge as always in Africa, is government implementation. The government in Africa plays an important role when it comes to agricultural politics, as they are the ones who have the mandate to act on such matters. The next section discusses the role politics plays and its impact on agriculture.

1.6 Agricultural Politics

'Politics and water resource management are two variables that cannot be avoided or overlooked, when it comes to managing water as a resource. In addition, these two factors are key in understanding how water as a resource can be managed. This is because the government plays a key role in the development of bureaucratic practices. It is, therefore, for this same reason why irrigation, politics and farming have continued to be an area of great scholarly interest' (Barnes, 2017: 149). According to Barnes (2017) this has demonstrated how water infrastructures are embedded in webs of social, political and power relations. Newell et al. (2017: 11) argue 'that by exploring power dynamics in Kenya, one can be made to quickly understand the numerous tradeoffs that come between competing policy goals, climate change and agricultural policies that play out, in particular, local and national contexts.'

Hope came to producers, when the Reforms by the National Rainbow Coalition (NARC) government in 2003 gave life to some parastatals such as the Kenya Meat Commission (KMC) and the Kenya Co-operative

Creameries (KCC). However, this was not the case in later years as increased beauractic systems chased farmers away and favoured the private sector. Thus, this gave priority to expand irrigation programmes as articulated in the Second Medium Term Plan (SMTP) of 2013-2017 (Recha, 2018: 17). Although, a wide spectrum of polices (SMTP 2013-2017) have been brought forth in the agricultural sector Recha (2018) argues that their continues to be absence of coordination among implementing ministries and the absence of common objectives. This is because the agricultural sector has suffered from outdated legal and regulatory frameworks and policies that have served to constrain agricultural development, trade and effective competition. In efforts to address these policies and institutional challenges in the agricultural sector, the government of Kenya has adopted the 'systems approach'. Thus, synergistic functions will lead to the attainment of set objectives, that will have the agricultural sector to be seen as part of a component.

It is important to understand the combination of relationships of power that operate among and between institutions active in agriculture. This is because their vertical relationships of power enable one to explain the patterns of governance that can be observed between the different policy players Newell et al. (2017). Furthermore, as Majiwa (2018: 3) claimed, the 'importance of rice as a crop, remains a political commodity since in most cases the government is a major player in the development of the rice infrastructures such as irrigation facilities, input supplies, credit and market functions.'

'Irrigation development by small-scale farmers is complex as it involves many contextual challenges' (Lefore, 2019b: 164). For small-scale farmers to achieve high returns from their irrigated crops deep structural changes that address linkages between domestic and global markets need to be looked into. Furthermore, consultative decision-making opportunities need to be put in place so that political institutions are accountable. The reform of customary governance systems is also very important. In addition, large project designs cannot be adopted for small-scale irrigation farmers. This is because delicate balances need to be sustained to maintain agricultural sectors vibrancy and local innovations.

1.7 Ecofeminism

Women and nature share a close relationship (Devi & Kumar, 2013). Women interact with nature in a variety of ways in every society. As such the ubiquity and universality of this phenomenon is profound and unequivocal. The issues at stake are whether women enjoy *de facto* or *de jure* rights regarding access, exploration and exploitation of natural resources; and whether they have the capacity and capability to interact with these natural resources in their domains meaningfully (Saka & Adebiyi, 2021).

Women, including young girls, constitute one of the most vulnerable groups in Africa. In addition, as in the rest of the world, they play crucial roles as domestic workers. Furthermore, they assert that women are depicted as being 'saddled' with home and domestic chores, while men are viewed as enjoying the exhilaration of life in the 'outside' world. Mostly, they view societal roles of men and women as being different. As mothers, wives and housemaids, women perform vital roles such as childbearing/nurturing, homemaking and other domestic services which are crucial to the family's well-being as a social unit (Amusan & Olutola, 2017).

Amusan and Olutola (2017) argue that women work very long hours in agriculture and that their role in agriculture is primarily subsistence and is a means to ensure household survival, supplementary to whatever support they receive from the dominant male head of the family. 'Patriarchy,' according to them, is the prime obstacle to women's advancement and development. It creates obstacles for women to go forward in society and consolidates gender subordination in the context of men-women relations. The growing challenge to patriarchy, according to Menon (2018) is attributed to the work of feminist scholars and economists.

Feminism, according to Gaebl (2018) is a philosophy intended to promote gender equality. Gaebl posits that the Global North is dedicated to fixing the inequalities that women face worldwide, however, they are more often than not more challenging than beneficial. This is because they impose foreign cultural values and have a cultural superiority mindset, often on gender related issued in the Global South. Feminists from Western societies need to acquaint themselves with global gender issues within the context of those living in the Global South if any impact of change is to be seen. Poor women in developing countries are seen as being exploited from the largely neoliberal capitalist and globalization. Feminists from Western societies according to Gaebl (2018), must be devoted to the true version of women's rights advocacy, like feminist socialism that seeks to break the global market economy. They must grasp that women being suppressed are also able to exercise influence. These women are already fighting the global market economy through women led movements that are built on their work within their respective traditions, explicitly through eco-feminism. 'Eco-feminism is a neologism that was coined by a French feminist, Françoise d' Eaubonne, in her path-breaking book, 'Le Féminismeou la Mort' (Sharnappa, 2018: 3). According to Sharnappa, the fundamental intention of the book was to call upon women to save the planet. Later, eco-feminism evolved into a theory with contributions from various academic endeavours and activist movements against the destruction of the environment at the global scale, and with a focus on women and nature. Eaubonne's book argued that the liberation of nature could not be divorced from the struggle for women's liberation. Ecofeminism's main concern, as a sub-branch of feminism, is the struggle for womens liberation, and against the patriarchal world order. Ecofeminism connects women and nature and asserts that patriarchy has oppressed both. Hence, ecofeminism successfully combines the philosophies of feminism and environmentalism. Ecofeminists have several arguments that advocate for the connections between feminist thought and ecological concerns Tan (2017). For some time, there has been a general notion that concepts related to women, nature and animals were regarded as inferior. On the other hand, aspects related to men, masculinity, reason, mind and intellect were regarded as valuable and worthy. Tan (2017) views ecofeminism as a challenge to these false premises. Tan also contends that ecofeminism connects feminism, animals and the environment, and is based on the contention that women and animals have become the primary victims of environmental pollution.

'Women and nature share a deep and interconnected relationship, with gender playing a significant role in environmental decision-making. While critical choices about nature are often made by men, it is women who are disproportionately affected by these decisions' (Tan, 2017: 34). 'The oppression of women and the exploitation of the environment is highlighted by ecofeminism as a praxis-oriented theory. It does so by examining the nuanced connections between the two, grounded in empirical evidence, theoretical frameworks, and epistemological insights' (Sharnappa, 2018: 3). Tan (2017) outlines four fundamental principles of ecofeminism: First, it highlights the parallel between the oppression of women and the degradation of nature. Second, it stresses the need to understand the nature and roots of this connection. Third, it insists that feminist theory and activism must incorporate ecological awareness. And fourth, it argues that any effective solution to environmental issues must integrate feminist perspectives.

'Despite ecofeminist's differences in approach and perspective, they agree that gender and the environment are interconnected, and should be studied simultaneously to analyze them thoroughly' (Tan, 2017: 37). Suresh (2021) argues that State and non-state actors, regularly mention women as being identified by biological traits and are linked with the environmental world in some way. Women's subordination is often misrepresented within policy frameworks developed by decision-makers who make inflammatory statements that weaken the fight of women's recognition with the environment. Women's roles and identities are constrained by societal interpretations with regards to their physical appearance and are cast aside from taking an active part in society. Such arguments on physical appearance only seek to unite women and the environment which separates men and links women through their role of giving life through birth. This position disregards the diversity and unequal treatment that showcases women as a one-size-fits-all notion of womanhood. It also shows women as naturally affiliated environmental stewards compared to men. However, this is yet to be verified. It challenges women's right movement against male dominance in different forms and is not a simple distorted depiction of men versus women. Achuo, et al. (2022) cite Sturgeon (2009), who asserts feminist ecology (ecofeminism) as being a framework and praxis that connects the women's movement to environmental protection advocacy. Rao (2012) views this theory as a 'a set of guiding principles, a grassroots campaign, and an embodied way of life', that is able to give an examination of political structures that showcase the link between patriarchal dominance and environmental decline. The participation of an environmental factor in every feminist framework comes about as a result of a strong relationship that is present between the domination of women and nature, thus, ecofeminism is understood on an assumption. It is considered that any strategic plan for development will put women's agency and nature protection first.

It is widely renowned in available research that women are critical stakeholders in environment development as they are viewed to be responsible. When women's financial, leadership abilities and community involvement are not changed Niaghi (2019) argues the fight for environmental resilience may be unrealistic. Furthermore, research argues that environmental issues heavily rely on current social structures that impact on women's ability to be involved in social and economic spheres. Ghasemi et al. (2021) posits that enhancing women's agency can have great impact on environmental conservation. This view is supported by insights from Rao (2012) who asserts that women have a had a great impact in the participation of environmental protection and restoration.

1.8 Women and the Nature Connection

Most studies on women and the environment have revealed that in natural resource management, women are significant actors and major contributors to environmental rehabilitation and conservation, through their roles as farmers, and collectors of water and firewood. Women are closely connected to their natural environment and often suffer most directly from environmental problems. In Chinese history, the Earth is viewed as a powerful symbol of nature, such as Mother Earth, Earth God, Artemis in Greek mythology, and Mother River (the Yellow River). In addressing most of the current environmental issues such as climate change, ozone depletion, environmental pollution and environmental degradation it has been reported that women play a significant role. Because of their direct interaction with the environment, women have a deep knowledge of it. Women are therefore knowledgeable about the environment, they are also protective and caring towards the environment (Raimi et al., 2019).

Some ecofeminist perspectives suggest that women possess an inherent and deeper bond with the Earth than men, which is why they are often seen as the natural leaders in the movement to heal the planet through ecofeminism. This worldview emphasizes the interconnectedness of all life, with women playing a central role in caring not only for the environment but also for other marginalized and oppressed groups. However, this belief is not universally accepted among ecofeminists, some of whom reject the notion of a biologically determined link between women and nature. Gebreyohannes and David (2022), citing Merchant (1980: 75), note that 'the connection between women and nature is ancient, deeply rooted in culture, language, and historical narratives'. Shiva (1988: 37), writing from the Indian context, illustrates 'how women are intimately linked with nature both in imagination and in daily life.' She argues that women derive sustenance from the environment and are guided by nurturing, feminine principles that in turn sustain nature itself'. At its core, ecofeminism holds that the same ideology that justifies the domination of certain groups of people based on race, gender, or class also underlies the exploitation of the natural world (Gaard, 1993).

It highlights the parallel between environmental degradation and gender-based oppression, attributing both to patriarchal systems. Since the term was coined in 1974, scholars and activists have intensely explored the biological, anthropological, and socio-cultural ties between women and nature, further reinforcing the relevance and depth of the ecofeminist framework. Therefore, according to many ecofeminists, women cannot be liberated unless nature is liberated as well. As a result, ecofeminists urge the elimination of all forms of oppression (Gaard, 1993).

Gaard and Murphy (1998) also suggest that ecofeminism is about the interconnectedness of women and nature. It explores 'neocolonialism, colonialism, racism, oppression and the class exploitation of both women and nature.' Furthermore, it opposes and aims to destabilize the hierarchal binary divisions (such as self/other, culture/nature, man/woman, humans/animals, and white/non-white) where men rule as reason, subject, and master, colonizing women as nature, object, and enslaved person.

Ayuk Etang E. (2021) emphasizing the relationship that exists between women and nature or the natural environment, ecofeminists add a new dimension to feminist critical theory in a fascinating way. Feminism has focused on power and equality, but Ecofeminism is reclaiming space not through recognition by men, but by women, reaffirming themselves and their relationship with nature, as evident in women's natural ability as nurturers and givers of life. Lorentzen (2002) states that women and nature are connected conceptually, culturally and symbolically. These connections are articulated in several ways. Many agree with Ruether that Western cultures present ideas about the world in a hierarchical and dualistic manner that lives out how the world is organized. The claim is that dualist conceptual structures identify women as feminine while that of men as being more masculine. Nature, heaven/Earth, Dualisms such as man/woman, heaven/Earth, reason/emotion, mind/body, culture/nature, converge implying that men have an innate power over both nature and women.

Natural resources users (land, forest and water), are mostly women because they are responsible for gathering food, fuel, and fodder.

However, it has been observed that women spend most of their time working on farms to feed the household. This experience allows them to learn more nature (land, soil and trees). Women have become increasingly involved in agriculture because men are increasingly moving to urban areas to look for jobs, even though technological inputs have increased. This association of women and the land has made them much closer to the land and natural resources. Thus, this interaction has promoted a culture of respect for the Earth and the preservation of natural resources, thus ensuring that all the needs of the coming generation can be met. In addition, this secures access to the land and ensures better nutrition for the family. Women are responsible for protecting the environment's future. They prioritize protecting nature and have shown repeated interest in care for the environment.

The domination and exploitation of the Earth is seen in feminist perspectives as ecofeminism. This is why poorly resourced peoples and nature are at the heart of the ecofeminist movement. Women must play a vital role in natural resource management to achieve sustainable development. Furthermore, they must participate in natural resource stewardship. All women have a relationship with the environment. Women have a close relationship with nature, which makes perfect managers of the ecosystem (Mahour, 2016).

1.9 The Role of Women in Agriculture

Agriculture is a livelihood for approximately two-thirds of households in Africa. The sector contributes about 15 percent to the continent's GDP and provides income for over half of its population. Beyond economic contributions, increased agricultural productivity supports systematic change by enabling rural households to enhance education and skills and enabling labour to move to revenue generating industries, where few people are required to produce the same returns. Food production is also vital for food security, particularly as Africa's demographic is expected to rise four fold in the coming nine decades. Improved nutrition today will yield long-term benefits for subsequent generations. Therefore, channelling funds into the agricultural sector will increase output that is critical for enhancing financial growth and alleviating poverty. Sub-Saharan Africa is yet to achieve or meet the production levels that are present in other parts of the world, despite all these developments. A challenge to agricultural production is the ongoing gender disparity in the sector. To bridge this divide and unlock the full potential of African agriculture, it is critical to understand and address the gender-specific challenges that limit women farmers' productivity (Buehren, 2023).

Women's roles and positions in meeting the challenges of agricultural production and development are dominant and prominent. According to Ogunlela and Mukhtar (2009) their relevance and significance must be understood. (Forty five percent of the world's population, that is approximately 3.1 billion people live in rural areas, according to the United Nations (UN) Food and Agriculture Organization (FAO). A further 2.5 billion out of this figure view farming as a means of income generation and subsistence food provision. Within rural areas a quarter of the population are women. Thus, 1.73 billion women stay in rural areas, and a further 1.4 billion women have farming as a means of income generation. Another, 500 million, or just over a third are small scale farmers who are not land owners and get less than 5 percent of the agricultural supplies. Forty three percent of the work force in the Global South is composed of women, and half live in sub-Saharan Africa. Thus, 21.5 percent of women in this part of the world, compose of over a fifth of the working population in the Global South. Women make up forty three percent of the people who work in the Global South with 66 percent of them keeping animals. Women therefore make a great contribution to the agri-food systems in the Global South and their production in farms is critical in meeting the needs of the continent. However, limited access to land and resources hampers at least 500 million women (Glazebrook et al., 2020).

While conducting his study on the role of women in agriculture, Bryson noted that female farmers were the leading form of labour in crop output. Furthermore, in African villages, female farmers worked more hours than men. In addition, their labour was seen as having no value in communities where intensive farming systems were being done compared to extensive systems (Ogunlela & Mukhtar, 2009). Where disparity was present, it could be clarified in drivers like social systems. The African social structure seeks to place women at the center of farming systems, the African family and succession. It is a structure that is unique and not visible in other parts of the globe. Bryson concludes by stating that the failure to recognize/enhance female labour activities is contributing to current problems with food supply but these can be overcome most effectively by working with, rather than against, women. The merits of such collaboration cannot be overemphasized or quantified.

Women heavily assist in agricultural output, but have never been rewarded from support and technological advances that have been present. This has been attributed to cultural norms that seek to cast women aside, and not treat them equally. For example, in the FAOs 2023 report on the Status of Women in agrifood systems, women accounted for 66 percent of women employed compared with 60 percent of men's. Gender inequities is the main reason for this, rather than a lack of knowledge on agricultural programmes and projects. For example, on average women earn 18.4 percent less than men in agricultural employment; this translates to women receiving 82 cents for each dollar earned by men (FAO, 2023). In Africa, food is mostly produced by women and not men.

In Kenya, women make up between 42 percent and 65 percent of the agricultural workforce. This aligns with findings from the Luo community, where Pala observed that women's contributions to farming significantly surpassed those of men (Onyalo, 2019). Similar trends are seen across Africa. For example, women-led small farms make up approximately 80 percent of the nations food output in Ghana. Eighty seven percent of women in rural Tanzania working in agriculture, while in Zambia, women provide up to 80 percent of the labour for subsistence crop farming. Despite these vital roles, women still have major challenges in gaining access to knowledge across Africa. Deep-rooted socio-cultural norms limit their decision-making authority and reduce their opportunities to benefit from agricultural technologies. Many women have difficulty accessing extension services, which are often poorly suited to their needs or delivered at times that clash with household responsibilities. Furthermore, land access remains a major obstacle. Traditional and customary tenure systems reinforce gender inequality, and even where laws promoting gender equity exist, weak enforcement and poor implementation continue to hinder progress (Ingutia & Sumelius, 2022).

It is vital for women to fully embrace the emerging opportunities in African agriculture by gaining better access to key productive resources such as land and financial services, which are necessary to enhance their agricultural output. However, boosting productivity alone does not equate to full empowerment. Genuine empowerment requires supporting women in building entrepreneurial skills, entering specialized markets, and engaging in higher-value agricultural ventures. Additionally, agricultural research plays a pivotal role in shaping the future of African agriculture both by driving farm-level innovation and by addressing the persistent gender gap in the sector. Research institutions are increasingly focusing on projects that address real challenges faced by farmers and are offering innovative solutions. Including more women in agricultural research can introduce more gender-sensitive perspectives and lead to solutions that better reflect the realities of female farmers. Although progress remains limited, there have been notable gains. Between 2000 and 2014, the number of women working in agricultural research in Africa grew from under 9,000 to over 15,000 a 24 percent increase demonstrating a growing recognition of the importance of women's participation in research (Jimbira & Hathie, 2020).

1.9.1 Gender Inequality

The Beijing Declaration (UN, 1995) that emerged from the landmark 1995 Beijing Women's Conference included a focus on the advancement and empowerment of women, highlighting the need to challenge patriarchal and intersecting structures that subordinate women in society and create gender inequalities. While gender inequalities related to rights, resources and responsibilities in the agricultural sector have been well documented, less well documented are gaps in empowerment and agency between men and women, owing to the lack of individual-level data and the lack of consensus (Quisumbing et al., 2023).

The behaviours, activities, characteristics, and opportunities that societies deem appropriate for girls and boys and for women and men is used to encompass the word, gender. Although influenced by biological sex, gender is a distinct social construct that shapes how individuals identify and how society expects them to behave. It forms part of the broader framework through which individuals understand what it means to be a man or a woman. On the other hand, inequality, refers to a state of imbalance or unfairness. As noted by Adeosun and Owolabi (2021), it often arises from both economic and social segregation. Structural barriers fuel this inequality by restricting upward mobility and concentrating access to resources among more privileged groups, while limiting opportunities for marginalized populations, particularly women.

Gender equality has remained a central focus of global development efforts for decades, yet achieving it remains an ongoing challenge. Despite increased global awareness and research attention, progress has been slow and uneven. Notable disparities still exist especially in areas such as access to financial assets and the disproportionate burden of unpaid domestic work shouldered by women. For instance, only 60 percent of countries provide equal access to financial services for women, and just 42 percent grant them equal land ownership rights. Addressing these persistent inequalities will require tailored, innovative approaches that consider both overt and systemic forms of gender-based discrimination. In addition, women spend, on average, twice as much time on housework and other unpaid activities as men (Njuki et al., 2022).

Yu and Osabohien (2023) state that agriculture remains a crucial sector in many African countries, providing employment and food security for millions of people. Despite the pivotal role played by women in agriculture, their participation and empowerment in the sector have often been limited due to cultural norms, policies, and inadequate resources.

In addition, according to Akpa et al. (2024) the connection between gender disparities and efforts toward inclusion is based on the foundation of three variables as outlined in academic research and policy documents; (i) Mainstreaming women into structured economic systems; (ii) The critical link between inclusive development and global sustainability outcomes; (iii) Shortcomings noted in existing literature.

Similarly, African women are often confined to the peripheries of the economy, participating primarily in sectors such as small-scale farming, petty trading, and domestic work areas that typically yield low economic output (Asongu & Odhiambo, 2020a; Efobi et al., 2018). The agricultural sector, in particular, is notably affected by gender disparities, a pattern also evident within the European Union. According to Akpa et al. (2024), gender differences in agriculture have drawn increasing attention since the 19th century.

In rural areas, young women tend to shy away from agricultural careers, largely due to its male-dominated nature. Compared to men, women's involvement in agriculture is frequently undervalued or overlooked. Socially, Gali'e et al. (2016) assert that in agriculture women's contributions are often ignored. Several factors affect women's participation in the sector, including limited access to financial services (Huyer, 2016), insufficient agricultural extension support, and the broader implications of their involvement on healthy diets and safety (Rao et al., 2019; Sraboni & Quisumbing, 2018a). Malnutrition is seen to persist due to inequalities in agriculture due to persistent gender inequalities (Malapit & Quisumbing, 2015). Food insecurity is persisted among members of family farms due to significant gender disparities (Tibesigwa & Visser,

2016). Improving women's access to education and secure land tenure requires addressing these gaps (Palacios-Lopez et al., 2017).

On the other hand, major challenges such as traditional patriarchal norms continue to pose hurdles (Meinzen-Dick et al., 2019). Men, generally have larger land holdings than women who tend to have smaller and less secure land holdings (Lambrecht et al., 2018). Research by Sraboni and Quisumbing (2018b) and Rao et al. (2019) have shown that better outcomes in diets can come as a result of empowering women in agriculture. Rao (2016) further notes that women typically have less access to essential agricultural inputs compared to men. Female farmers also face greater obstacles in obtaining financing for farm improvements (Huyer, 2016). Interestingly, Lawry et al. (2017) found that access to credit can be facilitated if women hold marriage certificates. Gender gaps are also evident in access to agricultural extension services, with women often receiving less support than men despite a greater need (Emmanuel et al., 2016; Hellin & Fisher, 2018). Women face persistent information asymmetries that further hinder their participation (Kansiime et al., 2019; Kristjanson et al., 2017). Studies also point to unequal access to agricultural knowledge (Zossou et al., 2017) and training opportunities (Mudege et al., 2017), which in turn restrict women's ability to adopt climate-resilient farming techniques (Nelson & Huyer, 2016). Gender disparities are evident in Agricultural Extension and Communication Management (AECM) practices, particularly concerning the employment status of young female farmers, as confirmed by Unay-Gailhard and Bojnec (2021).

Notably, these differences tend to diminish on part-time family farms but become more pronounced when examining access to environmental support. Promoting Gender Equality and Women's Empowerment (GEWE) in agriculture and food systems is crucial to achieving Sustainable Development Goal 5 (SDG 5) on gender equality. Advancing GEWE can lead to improved food security, enhanced nutrition for all, and a transformation of food systems into more equitable, resilient, and sustainable models (Lecoutere et al., 2023). Enaifoghe et al. (2021) emphasize that elevating women's roles globally remains a central focus in the pursuit of sustainable economic development and continues to be a priority on political agendas. Ensuring women's inclusion across all levels of government is essential for achieving comprehensive gender integration, which has far-reaching benefits for poverty alleviation, socioeconomic growth, and national development. Having women in political and economic decision-making is vital to, ensuring equal participation and leadership for women. Promoting equity within agri-food systems across other dimensions of social inequality is also fundamental to reducing within-country disparities, as outlined in SDG 10 (Reduced Inequalities) (Lecoutere et al., 2023).

1.9.2 Gender and Development (GAD)

The study of how women's roles have evolved during economic and social development begun during the beginning of Boserup's study in the 1970s (Singh, 2007). Singh highlights that the gender and development paradigm emerged as a response to the marginalization of women from economic development processes. Early development planners viewed women primarily as symbols of traditional culture and, as a result, perceived them as economic and developmental liabilities. Jaquette (2017) asserts that during the 1970s, development became both an academic field and a policy challenge.

This shift was driven by Cold War conflicts between East and West, as well as North-South debates over the causes of underdevelopment, which ignited significant discussions among feminists. These debates led to the establishment of a Women in Development (WID) office within the U.S. Agency for International Development (USAID), following the liberal egalitarian ideals of U.S. feminists. However, this approach faced opposition from various groups and was challenged by Marxist and dependency theorists, as well as cultural or difference feminists. The main short coming of (WID) is that it views women as distinct from men (Ts'ehlo & Musingati, 2024). WID was the feminist version of the modernist conception of progress promoted in developing world contexts (Akpan-Obong, 2023). In Africa, WID emerged as a reaction to the general policy environment maintained by early colonial authorities and post-war development planners, who failed to acknowledge the economic contributions of women. Instead, women were almost exclusively identified in terms of their roles as wives and mothers. WID emphasized the need for women to play a greater role in the development process and advocated for the loss of customary beliefs and practices, which often relegated African women to household tasks (Ts'ehlo & Musingati, 2024).

Western feminist perspectives of WID employed an overarching description of women's social position, life experiences, and contributions while proposing one-size-fits-all approaches to women's issues in

Africa and the Global South. However, WID failed to address the unequal gender relations and power structures that underpin women's exclusion and subordination. The underlying assumption of WID was that women from the Global South were not already contributing to the economy. This perspective overlooked the critical role women play in family-based productive activities, as well as in informal economic and political activities. Latin American dependency theorists Frank (1967); Dos Santos (1998) argued that underdevelopment resulted directly from global inequalities created by the unequal relations of dependency between the Global North and the Global South. The same processes that enriched core countries simultaneously contributed to underdevelopment in peripheral nations. For example, the colonial extraction of resources and raw materials, coupled with financial and military dependency on the Global North, played a significant role in perpetuating underdevelopment. This system also failed to disrupt, or even interrogate, the patriarchal structure of gender relations that reinforces male privilege and power over women (Akpan-Obong, 2023). Rather, than promoting gender equality, WID inadvertently increased gender-based disparities reinforced through the exploitation of women as low-cost labour force and adding to their already burdensome workload in the unseen and unrecognized field of unpaid labour that sustains daily life and society.

WID was also criticized for its assumption that women's status would improve by entering 'productive employment,' implying that self-advancement required transitioning from the 'traditional' to the 'modern' sector. This assumption further suggested women's engagement in customary work roles across Africa and similar contexts has frequently acted as an obstacle to their self-empowerment. Additionally, WID's call for the integration of women into national economies was based on the flawed premise that they were not already participating in development. This perspective minimized the importance of women's contributions to household production, informal economic activities, and political engagement. WID initiatives contributed to the expansion of externally funded programs that promoted women's economic activities, however, these initiatives were not integrated into development programmes. Many of these projects originated through partnerships between donor agencies and state authorities, while women who were the key stakeholders remained marginal actors in the planning, monitoring, and evaluation processes. As a result, the WID strategy failed to improve women's agency, or enhance their ability to work in programmes. It also did not improve their ability to gain tangible and intangible supplies, which could have strengthened their role in societal management. Consequently, WID did not contribute to empowering women (Ts'ehlo & Musingati, 2024). Ultimately, WID was doomed to not succeed in the African context, which emphasizes a holistic perception of existence and the complementary nature of men and women.

1.9.3 *The Shift from WID to Gender and Development (GAD)*

By the late 1980s and early 1990s, the focus shifted from Women in Development (WID) to Gender and Development (GAD). As Marchand (2009) argues, gender and development is critical in the field of development studies because it bridges practice, policy, and theory while addressing the concerns of practitioners, academics and policymakers. Debates persist, as different perspectives exist within the Gender Equality and Development (GED) field, including Women and Development (WAD), WID (Women in Development) and, Development Alternatives for Women for a New Era (DAWN). The WID perspective is popular in policy circles, while the WAD perspective has gained traction among non-governmental organizations (Marchand, 2009). Gender and development, despite its widespread use, has not always aligned with the theoretical principles of these perspectives.

Parpart et al. (2000) emphasize that the Women and Development (WAD) framework highlights the specific and valuable contributions women make to development, particularly within African agricultural systems. 'Give credit where credit is due' was a slogan that captured and acknowledged WAD's recognition of women's unique skills, responsibilities, and objectives in development. Campaigns under this framework aimed to strengthen women's voices through independent local organizations and networks and influence policies.

It is also important to note that Gender and Development (GAD) built on the foundational efforts of WID and introduced key enhancements. Firstly, GAD shifted the focus to gender relations as a core component, stressing the importance of involving men in development efforts. Secondly, it called for gender mainstreaming, ensuring that gender considerations were integrated into all levels of project design and implementation. Lastly, GAD encouraged women to embrace and to articulate participatory development that would shape their own strategic gender priorities (Jaquette, 2017). According to Parpart et al. (2000), GAD reframed women as proactive agents in development rather than passive beneficiaries. It also critically examined how existing development policies and programs affected women's lives. As Jaquette (2017) points out, GAD prompted important shifts in bureaucratic practices to address gendered power dynamics more effectively at the grassroots level. Additionally, GAD strengthened the capacity of women's groups to push for legal reforms in areas such as family law and protections against gender-based violenc

1.10 Challenges for Women in Agriculture

Several studies have identified the challenges women face in their role in agriculture over the last few decades. The most prominent challenge remains the pervasive gender inequality that women encounter across all areas of life (Ugwu, 2019). Systematic forms of gender bias may be observed in:

1. Social norms, cultural ideologies, and prevailing attitudes that restrict women primarily to household roles,
2. Women's disproportionate share of household and economic labour imposes intense time constraints,
3. Legal frameworks and cultural practices that restrict women's ability to obtain credit, agricultural resources, jobs, education, and healthcare.

African women face many challenges in their role as farmers and do not receive the maximum returns for their efforts. The challenges hindering their efforts stem largely from socio-cultural and economic issues encountered both within the household and in the wider community. These challenges are deeply ingrained in social norms and have become a burden for women to tackle. Women play a significant role in agricultural production, and have over the years not been able to see the returns of their efforts until recently. This has been attributed to economic, social and cultural practices that erode the legal foundations supporting gender equality. Gender inequalities, rather than the absence of unawareness, explain the absence of women in farm related programs and projects (Ogunlela & Mukhtar, 2009). The next section will expand on some of the issues affecting women in the agricultural sector in Africa.

1.10.1 *Land Ownership*

Discriminatory land tenure systems significantly affect agricultural productivity particularly in food crop production and warrant close attention from policymakers and development planners (Adeniyi, 2010). This issue is especially pronounced in regions where the migrant labour system has led to a rise in female-headed households, many of whom lack authority and control over the land they cultivate. For instance, Kelkar and Govindnathan (2024) observe that in parts of India, women who legally hold land titles tend to have greater bargaining power within their households compared to those without any formal or informal land ownership. The situation is further compounded by rural credit systems that fail to recognize these gender disparities. As a result, women farmers are often excluded from accessing financial support, limiting their ability to contribute meaningfully to sustainable agricultural development.

The World Bank's 2023 report Gender and Agriculture in Sub-Saharan Africa: Review of Constraints and Effective Interventions, highlights land size and the quality of land as being critical factors influencing farm productivity, risk management, and investment capacity. Men consistently have access to productive land parcels, while women on the other hand, have smaller and less productive land. For example, in Uganda plots managed by women are 36 per cent smaller compared to men, while in northern Nigeria, women's plots are 42 percent smaller. Women also face lower levels of tenure security, which affects their willingness and ability to invest in land. Data from Ethiopia and Tanzania highlight that secure land tenure is strongly linked to higher productivity particularly for households headed by women.

Socio cultural norms-imposed on women, hamper their ability to access land, as traditions regarding land tenure mostly favour men. The introduction of Islamic religious systems in Northern Nigeria for instance, hindered women's rights to land despite the 1978 Land Use Act (Peterman et al., 2011; Aluko & Amidu, 2006). Under this law according to the World Development Report of 2012, a widow has less inheritance rights than her in-laws family, who would inherit her husband's assets in 61 percent of the cases. In Ghana according to Ragsdale et al. (2019) a husband is allowed to regain his wife land under customary law in return for an alternative piece of land. Thus, women did not invest in improving their land parcels as there was a lack of land tenure security (World Bank, 2023).

To understand how this debate about individual and communal land rights and how these rights affect secure tenure and gender relations, it is essential to explain the concept of property, land tenure, and secure tenure (Obeng-Odoom, 2012). Obeng-Odoom argues that property is seen as a broad term denoting tangible things, relationships, or rights to use and owning those things. When 'property' is qualified by 'real' as in 'real property' or 'landed property'. It becomes narrowly focused on land and the rights embodied therein. Property rights are therefore land rights. Galiè et al. (2015) argue that property is often associated with socially acknowledged and supported claims or rights based on law, customs and conventions. Galiè et al. emphasize the need to distinguish between 'property' (ownership) and 'access'.

They argue that the former entails 'the right to benefit from things' and the latter 'the ability to derive benefits from things', while 'access' has more to do with 'bundles of powers' (property with bundles of rights) and entails complex social relations that might change over time and space. Local definitions affect asset distribution in the household. Dancer (2017) argues that husbands and wives have equal rights in assets they have acquired together. However, when the land is passed from one generation to the next, the land is designated as ancestral or kinship land in large families. Therefore, it ties the land to certain gendered and intergenerational limits, and reduces the rights held by a person to distribute or transfer property or land. In Tanzania for example, within the agricultural regions of Kilimanjaro, Arusha and Kagera, the demand for land has grown and the traditional ties to family land strong as it is based on male line inheritance. Dancer notes this relationship with the land is changing due to socio-economic changes such as education, employment and the accelerated growth of cities which has created opportunities for buying and renting land. Muraoka et al. (2018) supported this notion and asserted that the land rental markets have historically been fragile in most of Africa, but this has changed post-2010. Furthermore, with a few exceptions, land rental markets are a significant way, if not the only way for enabling poor households to access land.

Thus, Ng'uono et al. (2014: 11) argue:

> land ownership or access to other productive resources and the organisation of agricultural production are influenced by cultural practices and traditions. For example, rules of land inheritance by lineage, gender or culturally determined characteristics give one core access to land. In some

communities land is owned by a tribe or kinship group where each family within the kinship group has the right to use as much land as they need to feed themselves.

However, inheritance patterns in Chiwambo village in South Eastern Tanzania, followed the female line, marriage practices were patrilocal, and inheritance patterns bilineal Dancer (2017). In patrilineal areas, some of the practices were different. Traditionally men inherit land from their clan, while women gain access through marital ties. In addition, male and female systems, traditional systems and land rights are tied to important events in life like marriage, divorce, separation, or the death of a spouse or father. Thus, land is transferred through a series of life events, for instance at the time of death of one's father or at the time of marriage. On the other hand, daughters may be allocated land if for instance, they are spinsters and are not married for a period of time or if there are no male heirs. Widows, if appointed administrators of their spouse's estate, may be granted access subject to their relationship with the male in-laws-brother/sons, and may be appointed to lead the family and land.

Ng'uono et al. (2014) argued that widows and unmarried women may be denied land and property rights in many of the rural areas. Thus, they must rely upon their fathers or brothers to provide them with a piece of land as family members. 'This leaves women struggling for their own subsistence. According to them, we must note that women's access to land is contingent upon their relationship to men and marital status' (Makhado & Pelizzo, 2016: 33). In some cultures, 'a man's possessions are inherited not by his wife or even his children, but by his mother's brothers' Ng'uono et al. (2014: 15).

Thus, the women have less drive to invest in or cultivate the land. In some communities, it is customary for a man to share his land between his heirs (sons) and exclude his spouse even when he is still alive. In other traditional communities, the land will not be shared when the kin (male farmer) dies. The parcel of land is returned to the kinship for reallocation. The use of customary laws in land allocation is every day in Africa. The disgruntling issue with these laws is that they often discriminate against women, preventing them from participating fully in agricultural production (Makhado & Pelizzo, 2016). Furthermore, according to Hausermann et al. (2018), cultural understandings view men as 'providers' and thus limit women's wealth accumulation and land access. In addition, according to Ng'uono et al. (2014: 9), 'despite, the roles of men and women being complementary; social and cultural expectations for women's unpaid household work creates additional pressure on their time, resulting in their low productivity and earnings.' Lawry et al. (2017) asserted that many farmers in developing countries hold customary rights as a highly secure measure in the context of local social arrangements. However, this may not be the case with regard to the legal status in the country's statutory property regimes. In such contexts, land allocated through customary arrangements is often legally classified as public land and placed under the management of government agencies. However, in cases involving commercial investments, these customary land tenure systems once provided secure land rights to generations of farming families have sometimes been overridden, resulting in the displacement of those families.

Boone (2019) explains that, in Africa, the drive to secure user rights gained significant traction in the 1990s. This shift partly emerged as a countermeasure against the market-driven dispossession of smallholder farmers and pastoralists. In agricultural regions, initiatives to formalize user rights aimed to safeguard small-scale farmers' land access particularly for women by offering legal protections. Boone argues that securing user rights can shield vulnerable groups from arbitrary land seizure by governments, powerful elites, or traditional authorities. It can also reduce disputes with neighbours and extended family, improve women's standing in land-related conflicts, and encourage long-term investment and increased productivity on family farms.

Moreover, clear and legally enforced land rights can support the development of land rental and leasing markets, enabling landholders and prospective farmers to engage in transactions more confidently. This, in turn, reduces risk and facilitates the adoption of improved agricultural technologies. Gender equity in land rights is explicitly guaranteed in Kenya's Constitution (Wily, 2018). It mandates the ending of discrimination based on gender in law, and traditional systems focused on assets such as land and other properties (Constitution, Article 60). Additionally, it includes provisions that protect matrimonial property and ensure spousal rights to occupy land in the event of a spouse's death (Government of Kenya, Constitution, Article 68).

The Land Act No. 6 of 2012 in Kenya mandates that compensation for the compulsory acquisition of community or private land must be paid not only to the affected individuals but also to any individual settled on a piece of land and their spouse(s) (Government of Kenya, CLA s. 107).

Additionally, the Community Land Act (CLA) ensures gender equality by requiring 'equal treatment of applications for women and men' (Government of Kenya, CLA s. 14 (4) (c) (i)). The law also recognizes that women who marry into a community cannot be excluded as members and retain their land rights unless they divorce and remarry in another community (Government of Kenya, CLA s. 30 (5). These legal provisions offer a basis for women to challenge unfair decisions made by community land committees. However, a significant shortcoming is the absence of specific requirements regarding women's representation on these committees. There is no stipulation for a minimum number of female members or a rule that women must make up at least one-third of the community members present to form a quorum during member assemblies.

Moreover, the law allows women to be allocated land independently from men. In Kenya, this has been catered for in article 27 and article 60 of the Kenyan constitution (2010). In contrast to what has been happening in Kenya 'Rwanda has been the exception in its approach to inheritance law reform' primarily influenced by the aftermath of the genocide. Rwanda's Constitution in 2003 sought to commit to gender equity principles for all Rwandans (Article 9), with prejudice being banned and subject to punishment by law (Article 11). In addition, equal rights of succession have already been established under laws governing marriage property, voluntary transfers, and inheritance Law of 1999 where all legally recognized children, under civil law, are entitled to equal inheritance rights, without any gender-based discrimination (Article 50). This provision is further upheld by the 2005 Organic Land Law, which bans discrimination based on gender or origin and affirms equal land rights for both spouses Article 4) (Dancer, 2017). The absence of data with regard to land ownership by women is astonishing, given the formidable role they play in achieving development objectives (Doss et al., 2015). Furthermore, data that is currently available is yet to be verified due to credibility.

This is because data available may represent a few examples, but may not represent the state as a whole. They assert that a single statistic masks differences that may be global, regional and within countries, and may have myriad considerations and challenges. They suggest that these matters need to be reviewed. Chigbu (2019) argues that policies developed with regard to women securing tenure rights, place women as a uniform group, which should not be the case. Policies need to be developed based on current experiences of women. On the other hand, the issues faced by women tend to be similar. In as much problems and issues faced by women are similar in the Global South, it is the grouping and lumping of women as one that is an issue, as it hampers the efforts of developing policies aimed at improving women's land rights in sub-Saharan Africa. Women's lack of agency in some communities with sub-Saharan Africa, may come about as a result of their role in household activities, and thus, places them in vulnerable positions as they are dependent on men. On the other hand, this may not be the case as seen in some communities in sub-Saharan Africa, where women are the providers. In this role women have agency and are empowered. It is therefore, important that disparities rather than only similarities are considered when mitigating development issues that relate to people's traditions (Chigbu, 2019). It is important that there is a clear understanding of asset ownership among women.

There needs to be a clear conceptualization of property ownership among women. Property rights must be explicitly established. This is because land documents can be used to ascertain who is the rightful landholder, but persons (landholders) may only have limited access to land in sub–Saharan Africa and the Global South (Doss et al., 2015). For instance, a land holder may be able to grow crops on a piece of land and bequeath the land to his offspring, but may not be able to dispose of the land, without the consent from his clan. Secondly, the sole data point applied implies that individuals own land. With further qualification, however, how jointly owned is classified is still being determined. Women's land tenure differs based on the land they own and its use. Women's land tenure is acquired from their family or traditional authority, is deemed more secure than those who have acquired property via informal land transactions (Chigbu, 2019). Specifically, it would be important to know how land is owned by married individuals in this type of research. There is also a lack of clarity regarding how communal land owned by clans, tribes, institutions, or government bodies rather than individuals is addressed in these studies. Lastly, any questions about the portions of land owned by women needs to be analysed along side that of men. This implies that if women possess only 2 percent of landholdings, the remaining 98 percent is assumed to be under male ownership. It is important that people know the amount of land available for sale. 'Two percent figure will be looked at in a different manner if men were to own only 4 percent of landholdings' (Doss et al., 2015: 404).

Kamade and Bahati (2019: 62) claim that 'individually and jointly held titles are potentially an imperative tool to enhance women's bargaining power within households. They cite Pakistan as an example, where land ownership has had a significant impact on women's empowerment and agricultural productivity.' Therefore, they recommend governments develop programs and policies to enhance women's productive roles. These programs will strengthen the property rights of rural women. According to the World Bank (2017), secure property rights constitute an essential ingredient for economic development. Consequently, policy-makers need to focus on measures that strengthen these rights. Additional gains may be achieved from remedying the distributional inequality in property rights between men and women. It therefore becomes essential to explore how land ownership affects other natural resources and how this relationship affects women in agriculture.

1.10.2 Water and Irrigation

Gender inequalities in water access vary enormously across many African regions. Studies that research the use of irrigation and the quantification of gender differentials in water access are few. A significant scholarship details the legal, cultural and practical aspects of women's relatively limited access to water. However, only two studies are related to the gendered water use in sub-Saharan Africa (Djurfeldt et al., 2018b). One covers the access to piped water in Limpopo and the other the distance to the main irrigation channel among forty-five farmers in Benin. Both studies found that women (either women farmers or female-headed households) were disadvantaged in their access to water. A notion that Sinyolo et al. (2018: 130) supported asserts that 'few studies have investigated the gender inequalities in water access among small-scale irrigators and its impact on household welfare.'

Women's control of water resources for drinking and irrigation is limited because societal values define men as the guardians of property and as heads of households. Men therefore make decisions for women regarding water resources in the public sphere. In addition, other factors such as ethnicity and socio-economic status influence women's access to water. Anecdotal evidence highlighted potential water access challenges among women, reporting that the gender and social position of women in communities has limited their access to irrigation water (Sinyolo et al., 2018). Notwithstanding this, control over natural resources is limited for the most women worldwide. This limited participation of women in decision-making can potentially fail government policies and programs (Le, 2018).

Women hardly have a role or voice in formal management and governance systems, especially at the community level and in the setting up of modern irrigation systems. Men dominate the decision-making processes relating to water allocation, even though they are not responsible for procuring water for food preparation and household production (Mitra & Rao, 2019). It is, therefore, important that the interconnections between women and the social institutions and scales of decision-making be in the form of informal networks and structures rather than formal ones. Thus, women's access to water for agricultural purposes should be improved, especially given their important role in food production. Ensuring women's access to adequate water supply is vital for addressing dimensions of agricultural production, poverty and gender inequities. Le (2018) asserts that water scarcity negatively impacts rural women and girls more than men. Women and girls are found to be burdened more through the collection of fresh water, which is attributed to gendered domestic roles. Furthermore, one's access to land ownership determines decision making about accessing water resources. Mitra and Rao (2019) claim that access to water for irrigation is usually aligned with landownership. In a pioneering study of land rights in India, Agarwal (1992) showed that women had various social and economic opportunities. They were primarily available to formal (mostly male) holders of land. Several field-based studies in Anthropology and Economics confirmed that women were efficient irrigators and productive farmers (Mishra, 2018). However, most canal irrigation systems allocated water to the official landowners, usually men. Land ownership patterns directly preclude many women from water rights. According to Van Koppen (2017: 9) 'land and water are intrinsically linked.' Van Koppen asserted that this underlines the importance of linking efforts to improve women's land rights with improving women's water infrastructure rights. Naturally, available water resources are location specific, and access to streams and other water sources requires a rite of passage over other riparian strips, which can be denied to women, as reported in Kenya.

'Women are often excluded from the management and planning decisions regarding water resources. They often must develop their strategies for dealing with water' (Kernecker et al., 2017: 2). They cite the example of Gambia where the arrival of pump irrigation made rice go from being a traditional woman's crop to a man's crop, and where men started to control both the farming and the revenues regarding the crop. This example highlights how rural development focusing on water resources specifically affects women, and their relationship to natural resources.

They showcase how women's adaptation to social networks and their use of landscape knowledge, ecosystems and natural resources contribute to their adaptation to limited water access in rural regions and landscapes. Parker et al. (2016: 6) affirmed that 'it is important to acknowledge the complex dynamics between men and women within households and on the farm. The distribution of work resources and the control across the homestead is fluid, and subject to constant renegotiation and flux.' In addition, these dynamics compound women's vulnerability to domestic and agricultural water insecurity (Sarkar et al., 2018). Johansson (2016: 14) argues that 'the inclusion of women, in many water projects appears only on the surface. Women are either briefly mentioned in water management descriptions or only given power on paper; and that even in some cases, husbands represent their wives in water committees.' Johansson (2016) contends that women are less recognized when they are viewed as irrigators, fishers or farmers.

'Access to water implicitly refers to water in the right quantity and quality, at the right moment and in the right place in short water available for use. To use water and derive the benefits from it, people need water in the appropriate quality, and at the appropriate site' (Van Koppen, 2016: 5).

To access water, communities often adapt in two main ways: by relocating to where water is available such as through permanent settlements, seasonal pastoralist movements, or flood-recession farming or by transporting water to where it is needed. Given water's vital role, Agwata et al. (2014) stresses the necessity of managing it sustainably. They advocate for integrated water resource management (IWRM), a holistic approach that coordinates the development and use of water and related resources to enhance social and economic welfare, and sustain ecological ecosystems and political dimensions. A successful IWRM approach requires inclusive participation, particularly from local communities and women. Quansah et al. (2019) emphasize that active involvement is crucial to achieving meaningful outcomes. Women's perspectives, especially in rural settings, are vital for addressing key issues in natural resource and water management.

1.10.3 Financial Resources and Assets

'The availability of finance is essential for sustaining the production of agricultural commodities' (Wulandari et al., 2017: 1). Moreover, the availability of finance contributes to increased income productivity improvements and efficiency. Lack of access to finance is a significant problem for the rural poor, especially in developing countries. Financial knowledge is an essential aspect of financial decision-making. Thus, a lack of financial knowledge, especially among farmers, can explain their low-income and slow economic growth. Ng'weno et al. (2018) contend that financial services are a core enabler for consumption smoothing, risk mitigation, self-employment, business growth, asset accumulation and wealth creation; and that access to formal banking services can help individuals and households make day-to-day transactions, plan for and pay recurring expenses as well as help finance small businesses, grow assets and safeguard savings against theft, and manage irregular cash flow as well as mitigate shocks from unforeseen expenses, and improve users' overall welfare.

Akter et al. (2017) assert that women invest as much as ten times more of their earnings than men in their family's well-being, in child health, education and nutrition. That is, they spend 90 percent of their earned income on education, healthcare and housing, on average, compared to 60 percent of what is spent by men. Thus, improving women's livelihood has a well documented 'multiplier effect,' because women's empowerment directly impacts agricultural productivity and household food security. Kaaria et al. (2016) highlighted that access to land and income can affect the participation of women in cooperative structures. For example, producer organizations require land for cultivation or cash as membership fees, which women need more. Furthermore, they noted that Meinzen-Dick and Zwarteveen's (1998) study in Sri Lanka established that only formal land owners can become members of water user associations. This was because men tend to have legal rights over land, thus making it hard for women to be members as they did not have formal legal rights. Ng'weno et al. (2018: 21) asserted that 'although credit accessibility is universally dismal, women may be worse affected because they are more likely to lack formal land titles that they can offer as collateral; and that they may be hampered by other bureaucratic documentation which demands good literacy.'

Rao (2016: 33) argued 'that the exclusion of women from formal financial institutions and 'chit funds' is well documented, and only highlights the difficulties women face when trying to access credit facilities in Kenya.' Rao (2016) viewed this as a consequence of women lacking property rights and limited participation in paid work. This was because their ability to access credit and insurance instruments depended on them providing proof of repayment via income streams or collateral, which a majority needed more. In addition, this may require them to travel long distances to formal institutions to lodge documents and meet with loan officers. However, Rao noted that this may only sometimes be the case, as women's involvement in informal institutions is more significant than men. In Kenya, more than 60 percent of the employees in informal enterprises are women (Rao, 2016). Financial inclusion can be genuinely transformative regarding women's productivity, livelihoods and economic empowerment and requires women's uptake and continued use of relevant financial products (Ng'weno et al., 2018). Rao (2016) asserts that this may only sometimes be the case as small microfinance institutions provide expensive credit with lower returns on savings which limits members' abilities to use financial instruments to increase their savings, investment and household incomes. Rao argues that small entrepreneurs in Kenya extensively used non-interest-bearing savings accounts, even though they had high withdrawal fees thus indicating the negative rates of return on their informal savings instruments. Despite an expansion of microfinance agencies targeting women and women's access to savings, credit still needs to be improved in rural and developing world contexts. This has come about due to their inability to build long-lasting linkages to financial institutions that can scale up access to funds and grow independently of non-governmental organizations (Rao, 2016). In addition, commercial financial institutions both public and private act in ways that further reinforce legal and social biases against women.

Ng'weno et al. (2018: 5) assert that a 9 percent gap remains between men and women in the usage of financial services in developing economies. This is because women face financial constraints that prevents them from accessing formal financial services. Paradoxically, even when financially included, women appear to get less benefit than their male counterparts. The participation of women in financial institutions can increase their ability to use credit and savings vehicles. However, whether they effectively control the allocation and distribution of these funds within the household remains. Group membership of women in microfinance can translate their participation into effective control.

According to Rao (2016), when female access to financial assets does not result in increased female ownership of productive assets and increased female income streams, the gaps between males and females can worsen. Historically, financial services interventions for small-scale agriculture have focused on loans for working capital to access non-labour inputs (such as seeds, fertilizers and pesticides) with some interventions aimed at longer-term, community-level projects such as irrigation or processing. With overall low credit access, it's hard to tell whether men and women have different credit needs in agriculture. Furthermore, men are more likely to make decisions about significant investments. Thus, it seems likely that men have more demand for larger, longer-term loans for irrigation or mechanization. It is therefore, essential to note that farming households have a definite demand for working capital loans. However, these working capital loans are primarily met in the informal sector, if at all. The informal financial sector is an unorganized sector consisting of money lenders, relatives, friends, neighbours, landlords, traders and groups of individuals who operate mainly in the rural setting and who seldom engage in lending money (Gbigbi, 2019). These informal financial groups are the primary providers of funds for promoting and developing small-scale farmers in rural areas. Furthermore, they are crucial in resource mobilization and allocation in developing economies. The lack of banking facilities is a significant obstacle to effective rural banking. It is claimed that in most rural areas where banking facilities are deficient, informal credit and savings markets are predominant and that farmers patronize moneylenders for both consumption and investment credit during land preparation and in the planting season. According to Omobitan et al. (2019), when farmers face all these challenges, it is interesting to analyze their financial choices to enable them to meet their agricultural investments and spending.

Thus, given these circumstances, informal credit sources are unquestionably the most popular source of finance for the rural and urban population because the formal credit sources have scared many food crop farmers due to the encumbrances. The World Bank global finance index report, according to Bizoza (2019: 52), shows 'that only 22 percent of women in low-income countries hold bank accounts in rural areas and that commercial banks tend to work only with large farmers already well positioned in the value chains.' In addition, the World Bank report estimates the gender gap in account ownership to be 9 percent in developing countries. It implies that women need help accessing loans or credits as they are not financially resilient, since their resilience depends on the security of their savings and their ability to have loans when needed.

According to Ng'weno et al. (2018), programmes that aim to transform women's productivity in agriculture and micro-enterprises have had limited impact. The reasons for this are complex and sometimes explicit but include the fact that financial services (alone) have little impact on productivity. Women may need more ability to benefit from the additional services (such as training and market access) that make financial services successful.

Products must be tailored to women's small, irregular incomes, close social networks and home responsibilities. Thus, according to them, technology offers the opportunity to reach women more cheaply and effectively. Furthermore, it will allow them access to products from the privacy of their mobile phones and provide them with more flexibility around the size and scale of the products. Inadequate access to productive resources limits women's livelihood options and exacerbates financial strain on women. Improving women's access to productive resources could unlock their potential, enhancing their efficiency and productivity. Therefore, the status of women farmers in accessing resources in different regions needs to improve (Sarkar et al., 2018). Access to affordable financial services, especially credit and insurance, enlarges livelihood opportunities and empowers people experiencing poverty to take charge of their lives. Such empowerment aids social and political stability.

Financial inclusion is critical for achieving inclusive growth which is required to ensure overall sustainable growth in a country. Evidence indicates that inclusive financial systems can empower poor households socially through the delivery of banking services at an affordable cost to the vast sections of disadvantaged and low-income groups who comprise marginal farmers, landless labourers, lessees, the self-employed, unorganized sector enterprises, urban slum dwellers, migrants, ethnic minorities, socially excluded groups, senior citizens and women (Omojolaibi et al., 2019). Financial management is essential for farmers in irrigation, and education is critical in helping farmers achieve their targets with regard to their return on investment. It therefore becomes important to understand, the impact of education and technology on farmers.

1.10.4 *Technology and Education*

Women's limited access to agricultural technology and education is an essential factor constraining women's agricultural productivity. Increasing technology adoption among women farmers has emerged as strategically closing the gendered productivity gap in agriculture promoting women's empowerment and advancing their broader welfare outcomes (Lefore et al., 2017a). According to them by helping poor women farmers use technologies they can produce more and maximize their returns on limited time, labour, land and capital. Udoh (2016: 120) argues that 'improved technologies serve no useful purpose except when they are given to farmers and the farmers in turn use them to achieve higher productivity.' Various technologies such as machines and tools, as well as improved plant varieties and animal breeds, fertilizers, pest control measures and innovative management techniques are crucial for maintaining and improving agricultural productivity and reducing drudgery in farming. However, in most cases, new technologies have been introduced taking only men into account. Thus, gender bias and patriarchal attitudes in institutions often reproduce assumptions that men are the only farmers who can cope with new technologies.

Consequently, knowledge, technology and training related to agriculture are generally less available to female farmers (FAO, 2019). Lefore et al. (2017a: 1) assert that 'men and women have different preferences and face different types of severities and constraints when it comes to adopting new technologies.' They believe technology adoption can be understood in Phase one: Awareness. Phase two: Try-out. Phase three: Continued adoption of Technology (Lefore et al., 2017). Each phase, according to them, presents challenges for women farmers.

They note that awareness is restricted by women's access to information and extension services to learn about technology and how to use it. Try-out is limited by access to and control over land, water, labour, inputs and other assets required to use technology. In addition, social learning and social capital reduce the perceived risks associated with the adoption and appropriateness of design, including affordability, cultural acceptability and sustainability for women's specific agricultural tasks and physical requirements. The last phase (i.e., continued adoption) is threatened by some of the same factors and other household members' appropriation of the technology. Thus, many of the constraints that technology seeks to alleviate are the same constraints that hamper adoption in the first place. According to Lefore et al. (2017a), it was reported that men preferred to irrigate cash crops like tomatoes, while women preferred to irrigate leafy vegetables that could be sold over a more extended period and over which they could retain control over income.

Furthermore, they noted that women in pump-owning households had less of a say over production decisions but used the pump for their plots. Given this, they argued that technology adoption should not be viewed as a goal, but rather as understanding technology's differentiated impact within the household and orienting technology promotion activities to advance specific development objectives. Development initiatives encouraging technology adoption can improve water productivity and must operate within this complex institutional landscape (Van Rooyen et al., 2017).

Agricultural development in sub-Saharan Africa has underestimated the crucial role of this institutional setting, and hence overestimated the value of technological investment. Gaining control over water comes at such a significant financial and institutional cost that many irrigation schemes are often referred to as socio-economic failures. Thus, irrigation systems are complex systems with many actors at various levels, scales and subsystems that interact and share knowledge in a coordinated way to be functional. Lefore et al. (2017a) assert that women in Ethiopia need to know about technologies because they cannot attend informational meetings. In addition, their domestic work prevented their attendance. They also noted that their husbands did not inform them when such meetings were scheduled. It is important that despite the technological advances in irrigation, crop varieties, agroforestry and fertilizers, most technologies do not reach female farmers as they receive no information about them (Sharma & Maheshwari, 2016). Furthermore, most rural farmers are illiterate and poor and do not adopt new technologies because they lack knowledge and cash. Wheeler et al. (2017: 759) present an example in Ghana. 'Water lifting technologies (small motorized pumps) were primarily accessible to better-off farmers, and the primary constraints on wider adoption were supply chains, the lack of access to financing, high operating costs, high price risks and a lack of institutional support'. Sharma and Maheshwari (2016) argued that when training opportunities for rural agricultural producers arise, they mostly go to men with formal education.

Therefore, we must expand and enhance agricultural production of principal workers (that is women). Some women opt out of training programmes because of cultural, religious or family pressures. Female farmer's low levels of education cause great concern (Hart & Aliber, 2015). Education is crucial to enable end users to adopt and optimally use new technologies. They asserted that the education levels of farmers wound need to be boosted, especially as farmers engage in more sophisticated input and output markets.

Education is considered one of the ways women can be empowered. Sulaiman (2018) noted that non-formal education will likely be more available to poor illiterate women in remote farming communities (Sulaiman, 2018). Adopting formal means of education is critical to strengthening women farmers' knowledge, skills and attitudes to help them cope with their multifaceted tasks. Sulaiman (2018: 39) argues that farming skills and related technology remain a man's domain and that recent analysis of curricula and materials reveals that despite visible guidelines and directions, the status quo reigns supreme, even though women's participation and contribution in farming and income generation plays a significant role for their family's survival. Namonje-Kapembwa and Chapoto (2016: 2) assert that, 'if men and women face the same constraints but their rate of adoption of new technologies is different, then it will be imperative to design technologies that best fit the needs of each group. On the other hand, if the difference in the rate of adoption is because men and women face different constraints, then it will be essential to address this unequal access. These complementary inputs will thus promote broad-based agricultural growth among all small-scale farmers'. Kirui (2019: 4) asserts 'the importance of education and literacy in enabling farmer's access to information and knowledge regarding production technologies and market opportunities as it allows them to make better decisions.' He claims that studies have shown that farmers with just a few years of basic schooling were more likely to adopt and correctly apply agricultural innovations. Applying primary schooling to technology adoption allows a better understanding of how gender impacts small-scale technology adoption in households.

The distribution of rights to technology shapes whether technology adoption is sustained, as the anticipated distribution is likely to influence men and women's interest in trying out the new technology.

1.10.5 *Markets*

Amassing a gender perspective on value chains and markets assists in developing an integrated analysis of value chains from production, distribution, marketing and consumption. Gender perspectives promote the identification of gaps and challenges regarding women's access to markets. In a market, producers buy agricultural inputs, sell products and use income for non-agricultural products to buy food requirements for consumption. Access to agricultural markets by rural women can reduce poverty and malnutrition and ensure food security in rural communities. Although women play an essential role in agriculture as food producers, they are said to have limited access to land (Ngomane & Sebola, 2016).

In sustaining rural livelihoods, women play a vital role in agriculture. Despite facing limited access to essential productive resources like land and capital, they are heavily involved in food crop production. In many parts of Africa, women typically manage the income generated from semi-subsistence farming. On the other hand, as circumstances evolve, gender dynamics in agriculture shift. When agriculture becomes more commercialized, women often lose control, as cash crops are typically managed by men. Additionally, commercialization is frequently accompanied by the introduction of new technologies, which can further marginalize women's roles in agricultural activities (Fischer & Qaim, 2012).

In Tanzania, farmer marketing groups led by women often struggle more than their male counterparts to access markets, an issue that Djurfeldt et al. (2018a) suggest may stem from women's limited access to natural resources. Markets are frequently dominated by men, who tend to design access frameworks that serve their own interests, often unintentionally sidelining women. A key challenge is that women's restricted land access makes it difficult for them to meet the scale and quality demands of formal markets, which typically favour well-resourced farmers (Ngomane & Sebola, 2016). While discrimination against women in market access may not always be overt, it often leads to exclusionary outcomes. Barriers such as high membership fees and the long distances to urban markets disproportionately affect women, who must juggle economic responsibilities with unpaid domestic duties like collecting water, gathering firewood, and participating in community events. Consequently, women tend to favour local markets, which are more accessible but less profitable. These local markets often result in intense competition and lower earnings, forcing women to sell their goods at reduced prices. Rodgers (2019) adds that gender bias in product markets further exacerbates this disparity. Women typically grow subsistence crops, while men on the other hand, are likely to cultivate high-value crops or crops for export. Social norms often limit women's ability to scale up crop production as they are primarily responsible for household food provision. Factors such as limited land ownership, smaller plot sizes, and lack of access to extension services and climate adaptation tools further constrain women's ability to engage in more profitable, market-oriented agriculture. Women also need information on new technologies and systems to increase their production. Technology helps women access markets and network within value chains. Women need help moving from subsistence farmers to commercial farming because they still depend on old farming techniques and out-dated technology which do not help them increase their yields at the same level and speed as their male counterparts. Women thus need to be empowered with new farming skills while taking cognisance of their dual role of performing both productive and reproductive roles (Ngomane & Sebola, 2016).

Nugroho (2021) asserts that as individuals, farmers must improve their capabilities through training or obtaining a higher level of education. Farmers need training in effectively using mobile phones or other modern ICTs to help them obtain useful agricultural and market information. Doing so can help smallholder farmers enhance their ability to interact with the market and ensure that they become active market participants.

1.10.6 Labour

African women lack access to labour according to the World Bank 2023 report. The report asserts that women's inferior access to labour is a barrier to closing the gender gap in agricultural productivity across all six countries on which the report focuses. A joint World Bank and UN (2015) study recently found that women's lower access to male family labour accounts for 97 percent of the gender productivity gap in Tanzania and 45 percent of the gap in Malawi. In addition, an analysis on the determinants of the gender gap in agricultural productivity in Nigeria revealed that the gap was traceable to labour market imperfections which tend to be biassed against women (Olakojo, 2017). Women lack access to labour for some reasons. Firstly, female farmers tend to live in households with fewer members, so they have a smaller pool of household labour to draw upon. In northern Nigeria, Oseni et al. (2015) found that female plot managers live in households that have an average of 1.26 male household members, compared to 1.94 in the households of male plot managers.

Some of the reasons why women live in smaller households are related to widowhood, migration and divorce. For example, in Tanzania, 67 percent of sole female plot managers are divorced (World Bank & UN, 2015). Moreover, in some countries, even after controlling for the number of household members, female farmers deploy fewer male household labourers to their plots and those labourers who work on female-managed plots tend to be less productive than those used on male-managed plots. Men are less inclined to work hard for a female boss, women have less control over the allocation of household labour and women have less time to supervise their labourers effectively. In the case of women's lower returns to hired (i.e., non-household) labour, a key factor may be that women are less able than men to afford the most productive workers. Within 23 households, priority is given to male plot managers first, so female plot managers are left with less and less productive labour (Pierotti et al., 2022).

We must note that for women farming in male-headed households, the amount and quality of labour that they can access for their farm plots is, in part, a function of intrahousehold negotiations. Decades of research from a variety of contexts in sub-Saharan Africa have found that households must collectively maximize agricultural production as one unit. Udry (1996), for example, finds that in Burkina Faso, output on women's plots is about 20 percent lower than output on men's plots within the same household, planting the same crop in the same year. These gendered productivity differentials are visible in the differences in fertilizer application and also importantly, labour (i.e., household male labour, children's labour and non-household labour are all used more intensively on men's plots) (Pierotti et al., 2022).

Typically, women have less access to male family labour in cases of divorce, separation and widowhood. For women smallholders, income constraints limit their capacity to hire male wage labour. In the United Republic of Tanzania, the lack of male family labour explains virtually the entire gap in agricultural productivity, while in Ethiopia and Malawi it accounts for about 45 percent of the agricultural productivity gap (Rodgers, 2019).

Research on gender differences in agricultural productivity draws primarily on collective or non-cooperative bargaining models in which resources are pooled together, and individuals make separate decisions about consumption and investment. In this model, the 'conjugal contract' is important for men to have a right to ask for women to come and work on their farms, and the same applies to the women with regard to have men work on their land. This example, is critical in explaining the negotiating abilities of individuals as well as their behaviour within the home, and outside the home. For example, in households where individuals cooperate, people may not fully influence the manner in which work within the home is done, but may on the other hand choose to showcase their roles as the head of the home, even when they are not bringing any economic value. Thus, those within the home accept inequalities, when resources are being distributed in times of difficulty. The use of barter trade within households is key for this research, where many women work together in farms with their spouses on huge tracts of land, where they are able to gain, but are not able to decide on what happens on the farm output, but still are able to manage an additional smaller plot which is theirs, where they are able to manage the output. The produce that comes from women owned land is often sold to provide for household needs within the home; thus, women use the funds to provide clothing items for themselves, their relatives and perform other social responsibilities, schools' tuition and other household amenities. Within this setting, the choices about farm workers within the home should be grasped as part of a complex bargain of working and racing over the clock and supplies with the best returns mounting for people in some examples, and within the home in others. The results arise as a result of the bargains negotiated by economic returns but also social cultural norms Pierotti et al. (2022) (Fig. 1.1).

The theory above shows women have multiple activities and tasks to achieve, and thus have limited time to work on their lands. It shows the importance of evaluating the use of one's time by noting the different activities that limit experienced female farmers. By lining up each item and the time barrier, and marking its effect on labour time and output, the theory shows that policies that reduce difficulties may not affect an equal increase in labour time and output on women's farms. In the theory for instance, (see Fig. 1.1) the case of child rearing. Caretaking activities are directly associated with the allocation of managerial labour hours and influence the productivity of both managerial and hired labour. On the other hand, activities outside of the farm, that generate income during the day, and during hot and rainy seasons are arranged to manage labour time. Thus, mitigating challenges around childrearing, might be important, but

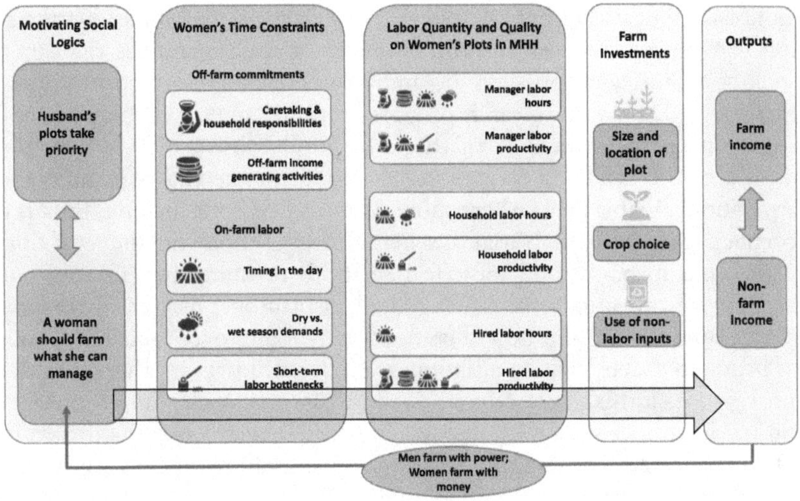

Fig. 1.1 Conceptual framework: How time constraints affect the quantity and quality of labour for women's farms (Adapted from Pierotti et al., 2022)

may not be able to create more working time as other interests on women may take precedence. The range of challenges women face, and how they intersect is critical in building mitigations that will improve their agricultural output. This theory showcases the need for social considerations around time use and labour allocation. The challenges women face are arranged in categories that meet social and traditional expectations around the right manner in which people spend their time. The effect of bargaining by different homes, reflects the need to maximise the output, and social structures. In addition, it is important to remember that arranging time use with social constraints may be good, but is subject to bargaining. Pierotti et al. (2022) show that examples of those changed from the traditions. For instance, Chineye shared how she compensated workers who took her place on her spouse's farm, by putting first her time for work on the farm. These examples, show the situations that make women to take up other options about their time.

Income from activities outside of the farm can be beneficial to women. Economists have linked the broadening of livelihoods to increased income, risk management and income stabilization (Barrett et al., 2001).

Feminist political economy scholarship criticizes the enforced fragmentation of work experiences and the instability that characterizes the varied livelihood strategies pursued by individuals lacking significant capital (Stevano, 2021). The research proposes a third angle on engaging rural women in activities outside of the farm; funds obtained from activities outside of the farm can assert intrahousehold bargaining over time and farm labour, Viable areas where more work can be done include; the need to check differences or variations between households in the way time is used and managed and the effect of the bargaining. In addition, one can explore the situational issues (social institutions, market opportunities, gender norms) and how the impact bargaining over time and labour for productive activities. In addition, a study that looks at other countries and regions, for example Nigeria to showcase the macro-micro linkages and the inequities within households in wider socio-economic differences (O'Laughlin, 2007). The theory will develop and examine future studies with regards to time use, agricultural labour, and gender inequities in agricultural output. Thus, when women have labour saving tools, they can reduce the time and effort they spend on the farm immensely, and open new ways for them to work in other economic opportunities that can give income for them and their families.

1.11 The Gender Gap in Agriculture?

One significant obstacle to the growth of the agricultural sector is the gender difference in agricultural production according to (Abdisa et al., 2024). In many sub-Saharan African countries where smallholder farmers predominate, women often have low agricultural output compared to their male counterparts. They note that women comprise over 50 percent of the agricultural labour force and contribute 45 to 40 percent, with the majority falling between 20 and 30 percent. Thus, when compared to their male counterparts their agricultural productivity is significantly lower. This has been attributed to deeply rooted socio-cultural norms and traditional gender roles that lead to unequal distribution of resources. In addition, gender based inequities in the ownership and management of economic resources continue to hinder agricultural production and undermine efforts to increase agricultural resilience and sustainability. A 2019 study conducted across 42 sub-Saharan African countries found a significant gender gap in land ownership, with a difference of 40 percentage points between men and women. On average, 31 percent of women in sub-Saharan Africa own land, compared to 69 percent of men (Wamboye, 2023). Wamboye further asserts that 57 percent of men in the Democratic Republic of Congo and 78 percent in Senegal do not own land, with the proportion being even higher for women ranging from 61 percent in Kenya to 92 percent in Senegal. He also notes that individual land ownership is limited for both men and women, though the figures for women are particularly abysmal. To close the gender gap, women need access to finance, financial services, seeds, fertilisers, and irrigation among others to increase their productivity. When women have access to technology, improved crop varieties and culturally appropriate machinery they are likely to reduce the gender productivity gaps between men and women-managed plots. Thus, to reduce gender gaps intervention strategies need to be evidence-based. Furthermore, these interventions should not only address only one constraint but a mirage of interventions as women need bundles of interventions to counter their challenges (Puskar et al., 2023).

The lower productivity of women in certain parts of Uganda and Nigeria has been attributed to the high cost of irrigation, fertilisers and herbicides (Julien, 2023). Julien asserts that the unequal access to inputs such as seeds, and fertilizer were not due to womens productivity and efficiency gaps, but rather the limitations on land rights for women, that were driven by social norms within Beninese society (Julien, 2023).

In Kenya women do not own land or moveable property and their land rights are dependent on them using their husbands or male family members land who own and control it (Hailemariam et al., 2024). Thus, Kenyan women are subordinate to men socially, economically and politically due to traditional gender roles that limit their opportunities. Sulaimanova and Jasoolov (2018) assert that addressing and reducing the gender gap in agriculture increases agricultural outputs by 10–20 percent thus, presenting a positive impact on household welfare, well-being and poverty reduction in low-income countries.

1.12 Closing the Gender Gap? Why It Is Important?

Gender inequality arises due to gaps in opportunities captured by unequal access to education, the legal system, finances or outcomes such as low female participation in employment, low wages and reduced political power (Ghosh & Ramanayake, 2018). Gowda (2018: 128) claims that 'in most societies, men's roles in agricultural activities are understood to be direct and clear. However, women's roles in agriculture are not recognized. Hence, the need to have a clear picture of women's participation in agriculture." "Women'. 'Women are leading the movement (impacts of globalization on the food system) in the hope of ensuring that food is produced sustainably, in both ecologically and socially sound ways. This is because they are constantly challenging the patriarchal paradigm, which is based on exclusion.' (Gordon, 2016: 58). Gowda (2018) asserts that although this is increasing it has been difficult to gain a clear picture of where and under what circumstances women particularly work on the farm. Women play important roles in helping their families in securing their food demand in the world. Women have had no place in agricultural history as it has been viewed as a male profession (Bozchelouie, 2018). In addition, the continuance of the profession, is seen to be passed from father to son. Bozchelouie asserts this social discrimination contributes to women in the agricultural sector being seen as invisible actors. He argues that women need to view themselves not as farmers, but as extra farm workers. In reality, according to Bozchelouie, females are 'only farmer's wives' and are to be seen as assistants. This is because men are seen as the main farm managers. This type of thinking has arisen from social norms about how women from rural areas should be viewed in their roles as housewives. His argument is that women are not biologically able to carry out farm work because of their physical strength. This therefore fuels an agricultural-related discourse which incorporates male views about female physique. 'Gender refers to relations between men and women and does not exclusively focus on women. In addition, the word 'gender' and 'women' is often used interchangeably,' according to Meinzen-Dick et al. (2014: 6). This is because gender roles are socially, rather than biologically determined, and as such are fluid and subject to change based on changing norms, resources, policies and contexts. Furthermore, gender roles are shaped by ideological, religious, ethnic, economic and cultural factors and are a key determinant of the distribution of responsibilities and resources between men and women (Meinzen-Dick et al., 2014). A problem that communities face is the lack of a common grasp of how women in farming communities should be viewed by their communities. This is because women are seen as helpers or assistants when it comes to farming and especially by those who have a strong influence in conducting studies, sharing of information and developing policies. Women encounter multiple challenges when farming, from lack of inputs and inability to access funding to support their increased production levels. All these resources are critical in production. Although the number of women farming has steadily increased, there still is an issue in state data, where women's work in the sector is viewed as invisible (Gowda, 2018). Shephard (2014) claims that examining gender as a variable that focuses on different experiences of women and men is a good way to study resource politics. Meinzen-Dick et al. (2014: 56) argue that 'explicitly incorporating gender analysis into discussions of agricultural productivity will expand the definitions of agricultural production. Furthermore, it will provide better insights into some of the constraints facing both male and female farmers.' In the Global South, women play a critical role in maintaining households and contributing heavily to agricultural output. But, the inequities that exist between the two genders, make it hard for women to succeed because they do not have access to supplies that would make them produce more and reduce their additional burdens. Thus, it is not just women who are restrained, but also households, communities, and the country's economy.

Women in rural areas take on a wide range of roles and responsibilities, often distinct from those of men. In agriculture, they are actively involved in planting, weeding, harvesting, and tending to livestock. Alongside their farm work, they also serve as primary caregivers managing households, cooking, and looking after children and relatives. Despite significant efforts from both governments and international organizations, gender disparities remain pronounced, especially in developing regions (Gowda, 2018). In areas such as Africa, Asia, and the Pacific, women generally work at least 12 hours more per week than men. According to Gowda, any strategy aimed at making agriculture more productive and sustainable must take gender dynamics into account. In many low- and middle-income countries, women have yet to fully benefit from advancements in agricultural technologies and practices, mainly because extension services often overlook them. This has led women's advocacy groups to call for clearly defined targets that prioritize women in agricultural extension, business training, and access to technology. Meinzen-Dick et al. (2014: 55) argue that 'achieving gender equality begins with dismantling legal barriers that prevent women from owning property, running businesses, or making independent career choices.' Moreover, community-level initiatives that engage both men and women in recognizing the benefits of gender equality can play a transformative role. Such interventions are crucial for the development of inclusive agricultural policies that boost productivity, drive economic growth, and help reduce poverty.

1.13 Conclusion

Women play a critical role within the food ecosystem in the Global South. However, if we want to achieve[1] SDG goals one to five, we need to ensure that SDGs twelve to fifteen are incorporated in the fight to ensure women are part of SDG 10, where they face no inequalities in the food production ecosystem. This fight is not only done, by women alone, but the wider community as well. In addition, we need to ensure that developmental programmes that work with communities incorporate gender development programmes. Furthermore, these programmes should not only be set to tackle one aspect but a wider set of issues that reduce the inequalities of women in agriculture.

[1] https://sdgs.un.org/goals ([1] No poverty, [2] Zero Hunger, [3] Good health and Well being, [4] Quality Education, [5] Gender Equality).

CHAPTER 2

Patriarchy, Women and the Environment

2.1 Introduction

Females from the Global South need to uplift themselves and confront their subservience through their actions and decisions (Jones, 2017). 'Economic globalization has rendered both women and nature invisible in a systematic manner by showcasing the gendered political economy as dominating households and denying women and nature productivity' (Gordon, 2016: 52). 'The relationship women in the North have with capitalism is very different from women in the South. Women in South have fought to keep their resources and needs with the goal of building resistance to capitalistic projects, that seek to devour their way of life' (Gaebl, 2018: 4).

'The relationship between women and economics has long been characterized by mutual exclusion, particularly within theoretical frameworks' (De Luca, 2018: 36). Certain domains of power are shaped by economic ideologies that have been designed by and for men, effectively excluding women. For example, the term 'the feminization of poverty' was used to highlight the economic situation of women in the 1980s. Such terms have been used historically to showcase how women have had little or no agency in economic roles and how they have been absent from the influence of resource allocation. Furthermore, women have only been brought into these discussions, due to societal shifts in the perception and social construction of gender.

M. O. Jimoh and S. Narsiah, *Unveiling the Power of Women in Irrigation*, https://doi.org/10.1007/978-3-031-94174-0_2

'Poverty should not be assumed by one or any person due to economic stance. This is because what one culture may assume as poverty may not relate the same to another. Thus, poverty is not subsistence living. There are those who live on commodities not necessarily produced, consumed and distributed through the market' (Shiva, 2014: 98). For example, one may plant crops on one's farm, not necessarily for sale, but to meet the needs for one's family. However, if one has excess, one may decide to dispose this in the open market to earn extra income to meet the needs of one's family. Shiva's (2014: 76) argument is that 'nature and human needs are managed through market mechanisms that seek to ensure the maximization of profits and capital accumulation'. Another example can be seen with regard to using hay or straw for the roof instead of iron sheets, or the consumption of locally processed millet instead of the processed one. In short, using traditional or indigenous products does not necessarily mean one is poor, but rather seeks to reinforce the need to consume products from the capitalist market. Hence, reinforcing the need to remove 'poverty' as a project. Such actions having made assumptions about 'development' tend to be culturally biased, and seek to only highlight that the way other people live and work is improper and that the people are poor. Hence, this creates real material poverty, by diverting land and water resources away from sustenance needs, cash crop production, and food processing and thus deprives increasingly large numbers of people from satisfying their entitlements to food.

According to Shiva (2014: 101)

> 'the addition of more commodities and more cash means less life regarding nature through the denial of basic needs. She asserts that this brings about the devaluation of women because their work co-operates with nature's processes, and second, because their work satisfies needs and ensures sustenance in general.'

The patriarchal ideology that women should not be viewed as primary bread winners and their work rendered invisible, only seeks to reinforce the importance of women's role in farming and specifically subsistence farming and the challenges they face (Amusan & Olutola, 2017). Women's work within the agricultural sector and many other sectors is unrecognized by economists, as it does not fall within the economic sectors. This is not because too few women are working in the sector, but because many are. It is critical that women working within the agricultural sector are seen for the important role they play within the sector. Patriarchal capitalist economists need to include women's work inside and outside of the home, and not render their work as non-work just because it does not conform to the norm (Gordon, 2016). Neoliberal globalization tends to put women out of work as land is converted to industrial agriculture and other activities. 'Land can be saved from the exploitation of globalization' as argued by Gaebl (2018: 5), by recognizing that women's work is legitimate. Gaebl's, argument is that by having more women in crop production for both subsistence and domestic production, the need to import food will be discouraged and the need to have organic food products and goods will be encouraged. This in turn will reduce the financial drain of African resources and will initiate and draw more produce industries to the continent. Most importantly, this will be championed by women and hence, challenge the patriarchal dominion. This will then lead to food production that is organic and sustainably produced in ecological and socially sound ways and means. Women look at the wider picture and note that everyone is important in the cycle (Gordon, 2016).

2.2 The Commons—(Common Property Resources)

The central argument of the commons can be found in different books and different authors who explain the concept in different strategies. For example the prisoner's dilemma is the most widely used example, where to showcase the difficulty in co-operation between selfish human beings and rational. Where the players have all the information they need to decide on a strategy, but because they are unable to communicate to each other they come up with the least conducive strategy to carry them forward (Rierio, 2023). Olson (1965) was also able to do the same and found that in the same approach individuals within the same community can still fail to achieve a common understanding that would benefit the community at large. This was because some individuals within the group termed as 'free riders' have little or no motivation to help the group/community achieve its objective. The irony is that they would still benefit from the group action whether they take part or not hence, the paradox of the tragedy of the commons. On the other hand, Sabourin (2023) notes coordinated action does not necessarily go with the standard assumption of political economics. Sabourin criticized the 'first generation' models of rational decision-making frameworks (Olson, 1966) and their rationality and information actors. However, Ostrom used these examples to show that even with modelling and experiments, no outcome was guaranteed only when real actors were part of real situations. When real 'face-to-face' engagements or reciprocal knowledge happened there were possibilities that could lead to shared dedication, via the skill set gained by the actors, which in turn would allow the actors to gain knowledge through transformation of the regulations, and thus reduce the differences and allow for mutual advantage (Sabourin, 2023). Said (2023) argues that community participation reduces conflict in the use of community resources and increases a sense of ownership in local communities. The involvement of local people has, however, been criticized on the basis that local people are less involved in most local projects of community conservation and their needs have not been considered. This follows from the fact that most projects originated externally. Thus, conservation problems such as forest degradation and deforestation in most developing countries have not been solved. On the other hand, a new question arises that lingers on what can be defined as the political principle, that seeks to highlight the relationship between the people who in the community of the commons; looking through their common tasks, political and moral norms which frame their actions. The question arises not whether the individuals belong within this community, but the common burden which they share in preserving the resource they share. Thus, 'gendering the commons' can be a way in which women in communities are viewed as actors, and the role they play in resource management and the impact of the rules on their situation and on inequality (Périvier, 2023). Interestingly, gender was not a hot topic during Ostroms time. She, however, noted that governance rules would have an impact on women and men. Women can end up being excluded from the governance of the commons and thus strengthen patriarchy of organizations of the communities where women are, as the rules governing can end up excluding women. Furthermore, women's participation at the negotiating table does not mean that will be included in managing the resources (Périvier, 2023). Ostrom (1990) explained that a community is a social organization that can solve problems over natural resources that cannot be solved by individuals, states, or markets. She noted that, successful governance of common resources needs consideration of local rules. The rules that consider social norms and social learning processes, the biophysical environment of the resource system, and the rules in use clarify a common understanding of the governing processes and decision-making. Ostrom (1990) identified a set of eight (8) systematic principles as follows: 1. Clear marked borders; 2. Consistency between governance rules and local contexts; 3. Joint decision-making frameworks; 4. Tracking; 5. Scaled punishments; 6. Conflict Management Systems; 7. Minimal support for collective organization, and 8. Tiered governance systems.

2.3 Managing the Commons (Water Resources)

Water is given to man by the creator, therefore it belongs to 'no one'. It is for use by all for the good of all. Water as a natural resource is considered by rural women as the shared resource, or a 'res communis omnium' in legistered water policy. The Boran tribe say: 'Water is either a source that you "share in" as a member of a descent-based collectivity, or one that you "share out" to signify respect'. 'Natural resources belong to the living, the deceased, and the yet-to-be born'. Thus, belonging to a society should grant one access. However, access to resources is based on a person position within society. In the case of land, the activities carried out on the land are an example of one basis that can be used to gain access. Kinship is one of the most important ways in which women can gain rights to water as a common pool resource. Females who marry into their husbands' communities thus gain rights to water through marriage, however, their rights are not as strong as that of men. It is important to note that water as a resource is viewed as belonging to the living, the dead, and the yet to be born (Van Koppen, 2023). For example, in Burkina Faso rice paddy fields are considered precious gifts passed on from mother to son through in-law connections. On the other hand, men ignore their 'sisters' or 'women' in the paddy fields as they work more on rainfed husbandry, and their presence in such areas is believed to cause flooding (Van Koppen, 2023). To put into context men in most cases govern and organize groups that manage the irrigation systems within the farms. They are responsible for installing and piping and channeling the water to the farms. Men thus monopolize the water catchment areas up to those that are infested by snakes and around the regions that are high up in the mountains. An interesting outcome to note is that females join these programmes as delegates only and when they cannot agree on the water management then the land is confiscated and given to the (female) leader of the wetlands (Van Koppen, 2023). A similar example is also replicated in the East of Africa in WaLuguru of Tanzania in the matrilineal mountainous area. Where women and men work together in the construction and maintenance of irrigation systems. Female donation towards the program consists of nourishing the men working on the farm. The aim of this is 'to provide the energy and motivation for men to dig. Thus, making women to put in a lot more effort to look for additional funds to buy special food. Women would be embarrassed in the village if the men refused', as noted with the WaSukuma (Nkonya, 2006: 246). When scarcity of water looms, rules are put in place to ensure that all members of the community are able to have access to water. During water scarcity, the primary focus of multi-purpose infrastructure typically shifts to home use, animal watering, and accessibility of water to public infrastructures. However, these regulations are not always enforced and women, in particular, often face exclusion. Such rules, for example, can be 'first-come, first-served' or 'equal allocation to all' regardless of allocation may be based on financial contributions, plot size, or crop under irrigation. Van Koppen (2023) highlights instances where 34 percent of active female irrigators have challenged irrigators in the Pangani Basin lower basin systems. While (male) farmers took twice as many turns collecting water, many (female) farmers had not collected the first round. Frustrated, the (female) farmers began irrigating at the wee hours of the night, with men refusing to offer any form of compensation (Komakech, 2013). Similar forms of discrimination were documented in the Mkoji Basin by Sokile (2005). While the regulations in principle factor water collection turns, there is a societal order that has been adapted that follows the social status of those vulnerable in the community and consists of; widows, the poor, the disabled, female heads of households, married women farming independently, and men they are often ignored in practice. 'Men were found not to adhere to the rules, and were found dressing as "magicians" to scare women and vulnerable men irrigators from the field, when they were irrigating crops in the night. Women who were unable to go to the field faced additional costs of hiring men who could fight for water for them. In addition to the costs the male irrigators also intimidated them. Children were explicitly not allowed to the field unless with an adult' (Sokile, 2005: 182). The commons theory is seen as a candid example where capitalist practices can be shunned and the good of the resource and the community take on board, by those using the resource. However, in looking at the commons there is a need to look at the impact of gender on governing the commons. Men and women gain access to resources at different times and based on different settings. In addition, the impact of the rules governing their interaction with the resource or resources governs how they can make the resource productive for them or not? Thus, putting into question the economic and political order of capitalism, statism or patriarchy (Périvier, 2023).

2.4 Women and Land Rights

Women gain access to natural resources via relationships within their communities or between them and their families. This can be because they're in-laws, matrilineal, or patrilineal kinships. The value of environmental resources to females is linked in its provision of food, heat, water for drinking and washing (Sweetman & Ezpeleta, 2017). Families depend on the use of natural resources in helping them to provide food and other important commodities from the natural eco-system (lakes, rivers, forests, and fields). Like many environmental resources land and other resources are gendered. It is for this reason why Sweetman and Ezpleta assert that females lack real authority to sway decision-making about environmental resources and that women acquire access via the family relationships vis-a-vis marital rights. In the twenty-first century, women should not be left behind in their quest for property ownership. However, cultural norms have made it difficult for rural women to gain access to land, as their land tenure is tied to their relationship with men. In the past, this would have been acceptable, but at present this is seen as infringement of one's rights to ownership (Samandari, 2017). In one analogy by Sene (2018: 52) 'land is viewed as a woman who is welcoming a man in the act of procreation'. This means that men own land and that women cannot own land as they are owned by men, hence showcasing men's dominance over women. It also casts women aside as not being able to own land. This is because women gain ownership of land from men (husband, brother, or father). Thus, if they are widowed or divorced, they lose their rights to the land. In Africa, women are vulnerable as their link to land is affiliated to their link with those in their kinship. They thus do not have equal property rights as their rights are set with certain cultural norms of their communities and society (Samandari, 2017). 'Land is more than just a natural resource. It is a resource that confers identity' (Jacob, 2017: 76). 'It is an identity that carries with it culture, language, community, and social norms. When women and nature are put together, they can be able to break the patriarchal view that men have over them' (Lv & Wang, 2018: 1660). 'From an ecofeminist view many societies have historically placed many unequal burdens on women and men, leading to the domination of both environment and women by men. How? By giving men land ownership and casting women aside' (Samandari, 2017: 5). This is further proven by the fact that land titles are given in the name of men and not women, especially among female rural farmers according to Shepherd (2014). Such practices, only undermine the ability of female farmers to produce and to access the necessary loan and credit facilities that help them increase food production. In addition, such unfair practices go on to hinder their ability to protect their own interests. Shepherd goes on to share examples of China, India, and Philippines, where land documentation for women goes uncompleted and where women are largely excluded from land certificates, even though they hold the same rights as men. Interestingly, 'women have had more access to land in sub-Saharan Africa than anywhere else' according to Jacobs (2014: 185). He argues that land is viewed as a man's property and his wife as 'the first maid'. He attributes this line of thought to capitalistic colonialism which has undermined women's position in many traditional societies and reduced their agency. In addition, women do make any key decisions and have to rely on men (husbands or fathers) to make critical decisions for them with the patrilineal systems. 'Women conduct various household and non-household tasks in order to keep their households running and in doing so, men become dominant over them' (Samandari, 2017: 5). It is therefore important that land policies be gender responsive and aim to ensure that land ownership is not for one gender (men) and that both men and women can have access to land rights. In this gender equality will ensure that communities, both rural and urban have equal balanced land rights that not only favour one gender, but all genders. Samandari (2017) argues that the imbalance of gender comes as a result of men holding and making all decisions that affect the family. Thus, making land tenure a key issue in the struggle for rural gender equality. In the next section we explore how water as a natural resource is affected by patriarchal systems.

2.5 Her Water, Her World: Gender and Access to Water

Water is the most significant form of nutriment in life (Srivastava, 2017). It holds the productive, reproductive, symbolic, and synergetic values of human existence according to Jacob (2017). In addition, women and water are allied terms that mark the status of any society's dynamics.

Furthermore 'land, water and women are three interrelated concerns of development,' (Jacob, 2017: 76). Thus, the link between the degradation of land and that of women cannot be easily ignored. Water scarcity is at the center of the water crises, and that it is connected to poverty, power, and equality and not just the physical availability of water. Ecofeminists note that water as a resource needs to be protected for all and that it needs to be treated as a resource and not as a commodity, and that those who treat water as a commodity have failed to appreciate water in its natural form. The issue with water is one of governance in relation to equitable sharing for people and ecosystems (Magill & Benedict, 2019). The 'power, poverty, and equality' factors can be seen in water governance today. This is because it has become the natural resource that has been commodified and can be priced in the market. The question of this commodification is? Who then can be able to buy water? because marginalized communities cannot afford to pay the market price of water if it commercialized. The need for water by human beings is compelling (Magill & Benedict, 2019). Nkonya (2008) argues feminist environmentalists view women as not having an exclusive relationship with nature, as African women and men perform different duties and activities, furthermore that ecofeminism ignores the social construction of gender and nature. However, it is important that we understand that female roles vis-a-vis water are closely tied to the agricultural calendar. Women are thus structurally subordinated to men by patriarchal mechanisms that link access to key institutions via sex roles that are stereotypes and sexually divided labour (Sachs, 2014). Irrigation, according to Sachs, is a set of tasks assigned to men and is a gateway for them to access land, water, and cash. In this view ecofeminists thus fail to account for the differences among women, by class, race, and ethnicity. The connection between men and women and the environment must be understood in the context of power, property, and division of labour. Most water professionals and decision makers in water management and hydropower are men. The implication of this is that men have complete control of water as a resource as argued by Lebel (2019). Lebel's argument is that even if women still use water for irrigation, it is still a resource under men's control. He states 'the dominance of masculinity in irrigation management is so strongly normalized that it is not necessary to mention "men" alongside irrigators as everybody "knows" that agricultural water 'management is men's work'. In summary 'irrigation is made to appear 'normal' and absent of any gender issues. 'At community levels, women are natural resource defenders and are seen as linking their feminist struggles towards social justice and environmental struggles so as to have sustainable development' (Sweetman & Ezpeleta, 2017: 360). According to Perkins (2017: 3) 'the face of climate change, is moving and changing in the Global South and North'. She asserts that women who are largely leading this change are resisting the ongoing industrial depletion of natural resources and are working hard to reclaim the commons. This action, according to Perkins, has led the capitalist energy system to being seized and redirected towards common actors. This 'actually existing' movement of commoners is the result of the exploited taking over some of capitalist funds with a view of ensuring that universal needs are met. The understanding of ecofeminism resonates especially well in the works of Shiva and Mies (2014) who have sought to highlight the relevance of an alternative to capitalist patriarchy given that it has worsened the conditions for women as well as nature in the wake of globalization. Khanduja (2017) asserts that the ecofeminist theory has drawn attention to the links between development and gender by highlighting the fact that violence against women and nature are both built into the process of development (Khanduja, 2017). 'The Green Belt Movement, for example that was established by the late Nobel Laureate Professor Wangari Maathai, was focused on planting trees, environmental conservation and women's rights. In this response, the organization choose to focus on rural women who would plant trees for firewood and stop soil erosion and at the same time respond to the growing desertification that was growing in Kenya. Thus, the movement fought three fronts; government injustice, human rights and environmental safeguarding' (Das, 2016: 53). It is through the Green Belt Movement (GBM) of Kenya, that women have become increasingly mobilized and now grow seedlings and plant trees to prevent soil erosion. This shows the ultimate connection between 'mother nature' and women. The reasoning here is that women give birth, just like nature, and thus in order to restore nature to her unspoiled condition, it is mandatory to overthrow the patriarchal mentality which contributes to destruction and catastrophe (Samandari, 2017). 'Men destroy what women build up and men cause death where women give life' (Tan, 2017: 44). However, through all these efforts women continue to be negatively affected by the lack of access to agricultural resources. The 'gender gap' as asserted by Shepherd (2014) continues to grow and in order to reduce it, it is critical that women farmers are given the necessary support to grow them to the next tier of producers if we are to grow more and better food for our communities and children. The role that women play within the food eco-systems is critical. It is therefore important that we understand this role and find ways and means in which we can support women in protecting the environment (Mohapatra, 2019). This is because women are the most affected when land is devastated by ecological disasters. Furthermore, those who live in the Global South have to contend with corporate globalization and colonialism. We must, therefore, ensure that women are depicted as users and natural resource managers so as to have them secure nature for the coming generations (Lebel, 2019b). It is in this regard that we explore the role of capitalism and its impact on women and the environment.

2.6 Financial Resources and Assets

Environmental management operates within a capitalist system underpinned by a patriarchal logic (Phillips, 2020). Women and nature are exploited by capitalists (Giacomini et al., 2018). Ecofeminist action is a stance against this exploitation and affirms the commons, based on this understanding. The depletion of natural resources consumed does not take into account the possibility or impossibility of their renewal (Puleo, 2017).

Hence, in countries where the population lacks the political and economic power to face the destruction of the environment pillaging is limitless. Globalization represents the unification of authority within gendered capitalism, a primary concern for ecofeminists (Sydee & Beder, 2001). This phenomenon is examined through several lenses: the evolving concept of the state, the neo-colonization inherent in patriarchal capitalism, globalization's erosion of diversity, and the feminization of poverty. Sydee and Beder argue that ecofeminists are critical of both institutions and the state due to their central role in upholding the patriarchal economy. Furthermore, the state's evolving relationship with political participation is seen as fragmented and lacking a holistic perspective, often tied to nationalist violence and social disintegration. Regarding neo-colonization, they emphasize that the interaction between subsistence-based worldviews and capitalist systems reveals a pattern of domination over both women and nature, perpetuated by capitalist structures. Vandana Shiva argues that the Green Revolution did not feed the hungry, instead it increased hunger via technological systems that cast aside traditional farming methods practiced by female farmers and placed technology via multinationals to create economic dependency. This effect of homogenizing culture through commodification and market mechanisms is referred to this as 'McDonaldization'. It has been criticized as acting as a cover that chokes out differences. Inclusion is at the center of cultural and spiritual health for ecofeminists (Sydee & Beder, 2001). Women work on a casual and part time basis to earn a living, and due to this they earn little or no pay and without any job security. Thus, whether in the Global South or North women comprise of the majority of the poor. Sydee and Beder (2001) cite Mellor who considers sweat shop labour and prostitution tourism as examples of male-dominated capitalism that seeks to 'milk' women in its drive towards globalization. Thus, making it easy for women and nature to be viewed as easily disposable labour that can be disposed of, and making them live from hand to mouth. Ecofeminists view globalization as resulting from women's exploitation. On the other hand, Phillips (2020) argues that environmental resources and ecosystems will be saved by privatization and marketing, thus bringing them into capitalist systems. However, these trends act as evidence of the patriarchal ecosystem of capitalism. It offers organizations and practitioners a platform for presenting business cases that encourage settlement rather than promoting philosophical ideas that are strong. This, for example, can be seen in messages like 'unlock business value' using environmental strategies thus, creating an impression of a 'green economy' that seems to offer a justification for increasing sector activity. In this, nature is unable to communicate its value as it is always depicted by competition. This then creates a standpoint for interested groups and organizations, who then get swayed by important and powerful individuals who can take advantage of women and nature. Ecofeminists see the need for more radical action to challenge the capitalist system and the multiple systems that seek to destroy the environment. Thus, in their view there is a need to regularize free markets and to call for a fair system of distribution of resources as part of feminist eco-politics. This is thus one of the main reasons why capitalism is called into question as it seeks to cause devastation to non-humans in the world (Oksala, 2018).

2.7 Technology and Education

Gendered differences cannot be overlooked when it comes to the use of technology. This is because men and women often have unequal access to new technologies. Although most of the technologies may be gender neutral, project design and implementation may be biased towards one gender, often toward male farmers, thus hindering women's participation. In sub-Saharan Africa, effective application of agricultural technologies in production has strategic gender implications (Muriithi, 2018). In addition, even with the use of technology, female farmers roles may reduce even with the use of new farming technologies. Many major authors like Doss et al. (2001) have included many reviews on developing countries with regard to gender-related issues in agriculture technology adoptions. This is because men dominate science, technology, and business. Despite the importance of food to women's lives, decisions related to agriculture and food often rely on science and scientific data and contain little input from women. Furthermore, with increasing scientificity and commoditization which has been signalled by new food technologies, the values, knowledge, and experience of women are persistently sidelined and existing power structures further reinforced. The continuing devaluation and invisibility of women and other marginalized perspectives is readily apparent when considering how new food technologies are developed and regulated (Lee, 2018).

These reviews have focused on existing microeconomic studies on gender differences in use, access, and adoption of land and non-land agricultural inputs including the use of improved crop varieties, labour, fertilizer, credit, and extension services. Women comprise, on average 43 percent of the agricultural labour force in the Global South, with fifty percent in the continent of Africa. The use of mechanization in the continent is mostly for the production of commercial crops like rice, tea, and maize (Kahan et al., 2018). 'The manner in which technology is absorbed in different cultural contexts showcases the complexity cultural norms and values' (Muriithi, 2018: 5). This view is seen to be brought about by investments in labour-saving technologies. Furthermore, land tenure laws have encouraged a rapid degree of land fragmentation resulting in small-scale land areas being in more densely populated areas, making the use of mechanized equipment to become difficult in areas where labour is scarce. Kahan et al. (2018: 6) assert that 'the low levels have affected productivity and productive capacities, as farmers are unable to cultivate large areas of their land' (Kahan et al., 2018). This in turn affects women disproportionately as a result of low farm mechanization due to high labour drudgery. Bozchelouie (2018) argued that patriarchal ideology embraces the view that the aim of using machinery and agricultural technology is to remove women from the agricultural sector. They claimed, for instance, that a tractor is a symbol of men's power and a means of dominating women. This is because the benefits of custom hiring reduce the demands for initial investment, repair, and maintenance from farmers. In addition, this service can be lucrative for entrepreneurs who regard custom hire as a full-time business venture. To provide access to mechanization for small-scale farmers, the focus needs to be placed on promoting the hire service business. 'The use of mechanization can improve women's agricultural productivity and increase agricultural production' (Kahan et al., 2018: 10). Environmental education is seen as a means to ensure the population is not just aware about the impact of environmental degradation, but they are also equipped with the necessary skills to solve them. It is seen as a critical component for sustainable conservation. This culture is developed by changing community attitude and knowledge with regards to the environment so as to solve current and future issues (Mukoni et al., 2018).

Programmes that are not gender inclusive, and lack the incorporation of women and girls are bound to fail. The reason behind this is that men are mostly in charge of information and they rarely share this information with women and girls (Bozchelouie, 2018). Women have continuously been cast aside by men and their voices silenced when it comes to environmental education Mukoni et al. (2018). According to Amusan and Olutola (2017: 118) 'it is important to train and empower women through formal and informal education, in both subsistence farming and commercial agriculture. This strategy assists in achieving national food security, since agriculture is the mainstay of economies on the African continent'. 'Women along with other marginalized groups, are disregarded and are simply described as being "research subjects" in homogenous, ungendered ways' Quigley et al. (2017: 773). The training and empowerment of women can be facilitated through women's cooperatives or through state-sponsored initiatives which may be in the form of provision of extension services or through formal institutions like the Bank of Agriculture (BoA) and other microfinance banks which offer zero- or single-digit interest rates. In addition, these institutions can look at providing insurance cover to manage any associated risks with regard to the agricultural roles of women. Also, most women farmers in many African societies can be empowered as they are generally natural caretakers and administrators (Amusan & Olutola, 2017). It is hoped that the use of new technologies, can help women farmers adapt and adopt measures that will help them mitigate the effects of climate change and other ecological challenges. Furthermore, there is a need to also allocate agricultural extension officers who can help women farmers in understanding weather information that can help make the necessary decisions with regards to farming issues. Most importantly, the need to make the women friendly technologies at affordable prices as they are a marginalized group (Amusan & Olutola, 2017). Hosseinnezhad (2017) argued that Shiva (1989) considered environmental problems and crises in her writings, and concluded that humans have just one way in which they can save the environment and nature. According to Shiva, it is only by retrieving feminine principles and values as well as attaching respect for nature that this change can be accomplished. In addition, gender mainstreaming in all agricultural activities at all levels should be vigorously pursued. Women are the main group of people who can transfer eco-literacy and manage household consumption (Hosseinnezhad, 2017). How can these principles be achieved via political systems?

2.8 Agricultural Politics

'Environmental and social conditions require that feminism and environmentalism play a key role in environmental matters. Furthermore, discussions with regard to the environment are now changing and focussing on climate change and its impact on women' (Paleo, 2017: 27). 'This is because the management of natural resources requires equal participation in decision making levels in local, national, regional, as well as public institutions like parliament' (Caliyurt, 2016: 195). On the other hand, the number of women who work as professional agriculturalists, city planners, and oceanographers is still very low (Caliyurt, 2016). According to MacGregor (2017: 16) 'Environment scholars are developing ways in which the relationship between gender inequality and the environmental crisis will change politics, policy and practice in significant ways'. There is an important need for African governments, to legislate laws and policies that work towards protecting the environment and those living within the environment. The challenge arises when those same people drafting the laws and policies benefit from the destruction of the environment and encourage capitalistic opportunities that destroy the environment and the incomes of women farmers. The reason being that women are not viewed as being owners of the most valuable capitalist resource 'Land' and that since men are a majority of the legislators within law making halls, they can trample on both women and the land. An argument well summarized by The WoMin Collective (2017: 429) who argue that 'women are not viewed as being landowners by companies. This is because when it comes to negotiating land sales they are rarely consulted or compensated. Furthermore, community conversations engage men more than women. This raises the question of why African governments operating under a capitalist logic fail to empower women and their wider communities?' Agriculture is an important economic activity in Africa. How then can we close this gender gap and why is it important that we do so?

2.9 Conclusion

Natural resources are critical elements within the economic portfolio of businesses, in addition, they are also critical in our day- to -day living. The management of these resources also define the manner in which society, specifically women are impacted by their management. In today's society natural resource management is governed by governments and not the community. This, therefore, impacts heavily on communities and members living within the community, specifically women. Women bear the brunt in both social and economic outcomes, when they are unable to access natural resources, in either land or water. Furthermore, because of inaccessibility women end up suffering economic hardship as they strive to meet the needs of their families. Hence, the strong link as illustrated by Vandana Shiva on the impact of the Green Revolution on farmlands in Asia. Thus, natural resource management is a composite of not just economic management, but also nature. It is therefore important that we not only manage the resources that we have to ensure production of resources, but also to ensure availability of the resources, if not we may end up being just consumers who have nothing to depend on after a number of years when the resources are depleted.

CHAPTER 3

Women and the Challenges of Climate Change

3.1 Introduction

Environmental disasters driven by climate change cost massive sums of money and result in loss of lives and means of livelihoods (Doku & Phiri, 2024). Changes in climate patterns have caused increased food insecurity among the population in SSA most especially among adult women and dependent children; through erratic and uneven rainfall patterns. These severe conditions are disproportionately heightened in Africa a continent where irrigation take place in only four percent of the farms. If the development community and African governments can help women farmers produce weather resilient crops, such as millet and peanut, it will go a long way in helping reduce food insecurity among the most vulnerable. These foods are rich in protein and heathy fats and folate which is critical for child and fetal development. Doku and Phiri, note that if environmental action is not mitigated the cost of fighting environmental degradation will go up. Estimated figures currently stand at $69 trillion by 2100 if the threshold is crossed by 2°C.

M. O. Jimoh and S. Narsiah, *Unveiling the Power of Women in Irrigation*, https://doi.org/10.1007/978-3-031-94174-0_3

3.2 Women and Men's Adaptation to Climate Change

The need by both men and women to adjust to climate-related impacts is necessary if humanity is to survive this challenge. However, with regards to climate change one needs to note that there is a big disparity in the way men and women are impacted by climate change. The female populations face greater adversity compared to men when looking into climate change matters. The reason for this has been attributed to women limited access to education and other resources which hinders their ability to mitigate climate change. Climate change is likely to affect 70 percent of people living in low-income populations, who are mostly female. Females depend on natural resources to support their lives. They are often responsible for conducting household tasks. However, with climate extremities conducting such tasks becomes even more difficult and will require effective solutions to mitigate the effects (Sinaga & Harahap, 2023). Rural women face a high degree of negative challenges when it comes to climate change. These issues relate to increased household tasks, agricultural production and male out-migration for employment. That bear heavy consequences on family health and nutrition. In rural Australia, for example self-inflicted harm on men went up four-fold during the Millennium Drought, which decimated the country's farming sector. The same issue has also been facing India, with men committing suicide, due to crop failure and increased poverty. There has been a lack of comprehensive studies on the impact of climate change repercussions on men. Thus, the impact of climate change is born on both men and women and has significant impact on their health and well-being. Abdulai (2022) notes that the Women Deliver (2021) report found that due to climate change water and food scarcities exposed females to gender-based violence, as they walked many miles to get scarce natural resources. Female farmers face higher risks during natural disasters. Women often bear the brunt of this extreme climate events, as they navigate saving their children and belongings during times of floods. Women's cloths for example may not enable to safeguard themselves during raging flood waters and also give them the ability to maneuver raging flood waters. Hence, the reason why many women die during such calamities when they occur. The inability of women to be well prepared for such events, arises from their lack of information and their inability to access information, technology and resources, which then restricts their ability to learn and counter the effects of environmental disasters. In addition, gender norms, affect women's decision making abilities in the face of climate events. Women may not be able to persuade their partners to consent to updated environmental practices as they lack agency (Pyburn & Eerdewijk, 2021). For example, in Nigeria women are more concerned with the availability of subsistence crops, while men are more concerned about the effect of environmental change on their cash crops like tubers and legume crops (Christophe Béné & Devereux, 2023). Thus, there is an important need to actively mitigate the climate crises and its effects. The effects of environmental degradation are adverse and affect people and communities around the world differently. Cyclone Nargis, which ravaged Myanmar in 2008, forced many females into commercial sex work as a survival strategy; and dwindling farm yields due to climate change in Indonesia fueled female emigration to work overseas, exposing them to various risks. Against this background, eco-feminists and female environmentalists are deeply committed to promoting environmental health. This is because the inability of families to feed and accommodate their children due to extreme weather conditions resulted in early marriages of female children in Malawi, affecting an estimated 1.5 million girls; property losses and financial instability due to Australia's bushfires in 2009 led to an increase in domestic violence against wives and children. Gender disproportionately when compared with the West and sub-Saharan Africa tends to affect individuals differently. This is likely due to fragile economic and political states in the Global South, that are affected by environmental factors that cause low crop yields (Cameron et al., 2022).

3.3 Women and Water Accessibility

Sub-saharan Africa needs to prioritize heavy investment in its agricultural sector, so as to build resilient farming systems. These investments need to be both short term and long term, with the overall goal being that they are able to contribute substantially to agricultural production and practices in the continent. In addition, these investments will help develop the region to be more water secure, with a view of reducing environmental changes on the continent (Nhemachena et al., 2020). In Africa, the female gender faces a have a heavy burden of collecting water for their households. A task that takes numerous hours, depending on the source of water. This task is further hampered by climate change when fresh water resources dry out or are contaminated by flash floods thus reducing access to water for use household tasks (Sinaga & Harahap, 2023).

For example, Sultana (2011) study in Bangladesh, where groundwater was heavily contaminated by arsenic, significantly affected the ability of women from low social economic status from gaining access to clean water. Such examples showcase the consequence females face due to insufficient hygiene practices. In the face, of environmental matters, the lack of access to fresh water and contaminated water causes harm to females, and has significant consequences for both their health and agricultural production, as the amount of household chores and health problems increase. Arsenic exposure can result in skin lesions, swelling of limbs, ulcers and scabs just to mention a few. Single women are the most affected group as they are vulnerable to social exclusion, where they may experience bias and rejection linked to their physical appearance as they are yet to find suitable partners (Sinaga & Harahap, 2023).

3.4 Climate Smart Agriculture

Climate Smart Agriculture (CSA) has three objectives: one to scale up crop yields and to promote food security, two to improve the flexibility of agricultural production in response to a changing environment and three to lower carbon emissions from current numbers. CSA is a new concept that interfaces between science and policy making, as outlined by FAO in 2010 during The Hague Global Summit on Agriculture, Food Security, and Climate Change (FAO 2013). From these discussions two strategies were designed, one that looked at the scientific process and was started via international forums dedicated to CSA, and the other on policy making in 2014 which culminated in the creation of the 'global alliance for climate smart agriculture'. The importance of using climate smart agricultural technologies is that it helps in safeguarding farming resources while reducing the environmental impact (Venkatramanan et al., 2019). The adoption of climate smart farming is taken in different ways subject to both men and women's level of understanding of environmental matters and the type of knowledge they have. In addition, decision making with regards to the use of adopting climate smart practices is dependent on access to finance and other gender equality dimensions (Pyburn & Eerdewijk, 2021).

In India's Western Indo-Gangetic plains, women face a unique hurdle when it comes to adopting laser land-leveling (LLL). Gender norms often prevent them from directly interacting with men in public. This means a woman farmer can't simply walk up to a male farm owner or service provider, or even call him, to arrange for the service. Furthermore, traditional gender roles often dictate who does what work, further complicating access for women. This example, highlights how new technologies, like those for climate-smart agriculture (CSA), can inadvertently worsen existing inequalities if crucial questions about who's truly in control and who actually benefits are not asked (Pyburn & Eerdewijk, 2021). In other cases female farmers did not have the capacity to take up CSA until barriers of access to family labour and the lack of the right tools, transportation, and plating tools were addressed. The choice of female farmers to use CSA methods increased their work load and thus countered the benefits of increased farm output and income that they had gained. When new methods of farming are focused on male farmers alone, female farmers tend to be overlooked. Furthermore, female farmers tend to receive less information with regards to environmental matters compared to men, as seen in different study site locations located in Kenya and Uganda. On the other hand, female farmers in agro-ecological zones in Kenya tend to have weather related information while men have information related to droughts and flood (Gumucio et al., 2019). Climate smart interventions in the livestock sector assume that the impacts of environmental change are gender neutral, and thus neglect the gender power relations that affect men and women differently with regards to livestock rearing. These interventions, furthermore can increase labour investment and hence overlook the roles that women undertake in agricultural livestock rearing for example, like in the dairy sector. It is therefore important that the roles that women undertake in agricultural livestock rearing are recognized with the view of improving their access to credit services and training which will improve their outcomes. Thus, programmes should be designed to ensure gender differences are taken into consideration when designing gender inclusive programmes in the livestock sector. For example, it is rear for women to own cattle as they are primarily seen as belonging to the male head of the household, unlike chicken which are seen as more of a subsistence animal reared by women.

Gender dynamics thus, constrain women's ability to undertake livestock rearing and should therefore be considered when encouraging women to take up such programmes. In India, studies have shown that gendered acceptance for climate technologies need to align with female farmer's, as they represent a critical resource in farming. In addition, female farmers play a vital role in their roles as farmers and entrepreneurs. Thus, agricultural programmes should empower female farmers with information with the view of enabling them have the ability to make decisions with regards to crop planting. It is important to note that access alone does not guarantee change, if women's empowerment is challenged by societal expectations. Thus, programmes need to give women digital tools, that will provide environmental knowledge that will enable them be part of decision-making with regards to planting (Lau et al., 2021). If programmes targeting women are not supported by the community, there may be backlash from men. This focus on environmental adaptation can end up jeopardizing programmes due to lack of understandings of the objectives and the power dynamics within the community thus resulting in poor program implementation and gender in balances. In sub-Saharan Africa a programme focused on women's empowerment was widely welcomed within communities when communities were called upon to be part of the project and were gender inclusive (Lau et al., 2021). Alienated communities that are often led by women, are disproportionately affected by environmental degradation and climate change. Thus, it is important that alienated communities are protected from environmental injustices and have access to natural resources that ensure communities have access to clean water, air, and a healthy environment. Ecofeminism thus, highlights the challenges marginalized communities face. Ecofeminists call for a change of practices that seek to harm women and the environment. They call for a shift to sustainable agriculture and renewable energy resources with the aim of achieving better ecological and sociological well being for all. Ecofeminists note the importance of changing cultural narratives and worldviews that perpetuate the domination of women and nature with a view to promoting a culture that values equality, cooperation, diversity and competition as a crucial for sustainable development (Devi & Kumar, 2023).

3.5 Fighting Climate Change the African Way

Ninety-nine percent of resilience is very different for the 1 percent of humanity (Von Kotze & Walters, 2023). Living close to nature, allows one to form a bond with nature due to their mutual survival needs. For many people such as farmers, fishers, and other traditional individuals who are entwined with the environment, they must find ways in which they survive as they have developed a deep understanding of how they can learn from nature. Von Kotze and Walters give instances of nature inspired learning, where they mention Sicelo Mbatha who having grown up in an isolated village in South Africa's Kwa Zulu Natal province notes that:

> 'We humans have acquired much information, but have lost much wisdom. We have enough information to dig out the last lump of coal, to drain the last drop of oil, yet we lack the wisdom to protect our one and only home and source of life.' (Mbatha et al., 2021: 273)

Sicelo an outdoor guide has devoted time and energy to 'rekindle the wildness in all of us, to bring people into the presence of wildness and to help open our souls to its beauty, wisdom and infinite power to heal'. This awareness is viewed by Mbatha et al. (2021: 4) as acknowledging nature with all its existence. This sense of appreciation is drawn from his traditional culture, that views people as being one with nature. For Vandana Shiva, this understanding helps to draw us to the importance of saving and defending nature from those who seek to destroy it. Furthermore, the knowledge of this defence mechanisms is drawn from nature. 'I am one with them. I breathe the same air as the lion roaring for the moonrise and the dung beetle foraging underfoot' (Mbatha et al., 2021: 3). This unique form of learning, only goes to highlight the fragile teachings of nature, 'to receive teachings from more living beings than I can count from the fragile butterfly to the great, ponderous elephant' (Mbatha & Pitt, 2021: 7). He attributes the unique relationship of living with nature by citing John Mbiti (1969), who argues that if one is in harmony with nature, then one is assured of one's own wellbeing. However, if one is in disaccord with nature, then nature will show its dissatisfaction, through various natural disasters as can be seen in climate change disasters. This type of thinking may be seen as somehow undervalued; there may be some truth to it. African belief systems is anchored in nature; therefore, they understand the concepts related to eco-systems and environmental degradation (Wane & Chandler, 2002). Women have been held as custodians of land, and have passed this knowledge to the next generation via riddles, folklore, proverbs and stories. It is because of this that they have been taught to respect the earth. Chitando et al. (2022) argue that environmental change measures have always been a part of traditional ways that have been linked to religion and has been captured by Arewa (1998) who states:

> Mother Africa taught her children to respect the earth and all that dwell on her. She taught them the laws of nature and helped them to understand the cosmic rhythms. [African people], who created rituals to communicate with these forces, which they then deified and praised, knew the celestial realms and elements . . . Ancient Africans lived closely with nature and they realized that internal energies are also governed by [the four natural] elements [namely, water, earth, air and fire]. (p. 38–39)

Environmental scientists recognize culture and nature as being unique and linked and as being important in carrying out various spiritual components. African women, hold the relationship that they have with nature as being scared and one that needs respect. On the other hand those from the Global North seek to overpower and control nature. These same beliefs are repeated by women researchers on excursions related to women's traditional knowledge, where they find notions of faith and spirituality which has an effect on traditional peoples' options and dispositions, and how these impact their understanding of their environment (Wane & Chandler, 2002). In contrast, when other people initiate training on climate change, they tend to ignore women's knowledge related to farming and the local ecology, which then remains excluded or ignored. Wane and Chandler, continue to acknowledge that the ecological awareness by indigenous women has typically been viewed as not having value and therefore should not be taught as part of research or any form of formal or non-formal educational setting. Mainstream science may not view women's ways of knowing the environment, as key but it is important that institutions of higher learning incorporate the knowledge shared over the years by those who understand the earth best.

They argue that:

> Our educational institutions need to see their purpose not as training personnel for exploiting the Earth but as guiding students toward an intimate relationship with the Earth. For it is the planet itself that brings us into being, sustains us in life, and delights us with its wonders. . . . [The] future can exist only when we understand the universe as composed of subjects to be communed with, not as objects to be exploited. . . . Intimacy with the planet in its wonder and beauty and the full depth of its meaning is what enables an integral human relationship with the planet to function. . . . Nourishment of both the outer body and the inner spirit will be achieved in intimate association with each other or not at all. (pp. x–xi)

3.6 Platforms for Women to Address Climate Change

The female gender is closer to nature due to their household tasks of caring for the home and the family. So, having them as part of the team fighting climate change would improve their standards within the community. The need to have women and young people drive the call for climate action is driven by their sheer numbers in the number of poor, unemployed and underemployed. In addition, they suffer the most when it comes to social and economic deprivation. This is because they tend to use local resources that degrade the environment and have 'gender bias' effects. Thus, for visible change to occur, women must be at the heart of any environment and developmental programme. The reason for this is that they usually bear the burden that comes with negative effects of environmental degradation. Environmental challenges have a great economic impact. This impact is tied to rural women and their families, as they benefit from natural resources (Lekwauwa & Ijioma, 2012). Women's coping strategies and mechanisms has given them the ability to be able to live well and healthy. They have developed a close bond with the environment which has helped them develop better coping strategies and mechanisms than men.

Through this relationship women have learned how to farm extensively and have gained an understanding on how to relate to the natural environment. They can walk for long distances in search of water for their homes and animals, whereas men tend to migrate to cities leaving them behind with the children. They are thus prone to getting sick more than men. Young people and women have a critical role of balancing the natural environment with their needs for development. They have the ability to fight for the environment, because when the environment suffers, they too suffer. The growth within communities is the best measurement for development (Lekwauwa & Ijioma, 2012). For Durning (1990), 'True development puts first those that society puts last'. Women and the environment are linked, and are at the heart of environment conservation (Lekwauwa & Ijioma, 2012). Oladipo and Tseayo (1992), quote Thomas Sankara who noted that '...decision makers must concern themselves with women's issues. To devise development planning without the participation of women is like using four fingers when you have ten.' Juan Hoffmaister, a young activist from Costa Rica, condemns humanity's reckless actions, stating, 'Climate change is an issue of intergenerational justice. If we understand how our actions impact the planet, it would be criminal to continue as we are, knowing we are endangering future generations' (Lekwauwa & Ijioma, 2012: 303). The call to action for the climatic fight stands with the young people as they have the chance to create policies that will have a positive impact on the environment. They are critical in demanding change and being part of the change. The importance of having women and the youth as part of the team fighting climate change is to harness their potentials and empower them as they are part of the collective action plan and implementation (Lekwauwa & Ijioma, 2012). Working with young people early on is important in having them learn how to cope with environmental matters. This can be done via community programmes and engagements that seek to improve their knowledge via curricula and other integrated learning sessions. However, doing so requires the required actors to include gender mainstreaming as part of the process, with a view of having women, young people and the youth better equipped to mitigate climate change matters.

The Brookings Institution have noted that actors in the sector have focused on the following issues with regards to women and the climate: (1) they have focused on gender equality as a key theme in environmental areas such as farming and power; (2) improving women's agency when it comes to providing solutions related to climate change; and (3) increasing women's ability to be part of the green economy (Kwauk & Braga, 2017). Contrary to this, they note that one of the ways in which women can be championed as climate activists is by promoting girls' reproductive rights. Family planning is seen by environmental economists and researchers as a better way to reduce greenhouses gases than other means. The argument here is that educated women are more willing to use family planning methods than those who are uneducated. For them, this would result in the reduction of carbon emission by $4.50 per ton compared to $5 per ton from the agricultural sector and $20 for low carbon strategies (Kwauk & Braga, 2017). Population growth is thus used to highlight environmental issues, because of poor family planning by poor women in the Global South as noted by Jade Sasser who is cited by Naples (2020). Such notions ignore, the other factors that are known to assist in environmental degradation and have historically come from the West. During the Climate Change Conference (COP 17) in Copenhagen, Zhao Baige, noted that 'dealing with climate change is not simply an issue of CO2 emission reduction but a comprehensive challenge involving political, economic, social, cultural and ecological issues, and the population concern fits right into the picture' (Sasser, 2016: 57). This ends up creating a story line that tries to explain complex environmental issues. This approach is used to avoid the real issues affecting environmental degradation in the Global South, while encouraging population reduction in the south (Sasser, 2016). The use of such strategies is just one of the ways in which men try to control women with a view to strengthening gender inequality (Monbiot, 2015). Women need to be part of the environmental actors who are mitigating the effects of climate change. They can do this by (1) sharing the knowledge they have gained working and living within the environment (2) by ensuring that they are part of the leadership teams making decisions with regard to environmental changes (Kwauk & Braga, 2017). The reason for this is that tools used to fight climate change are built by powerful people in the Global North who use them to deny women in the Global South the ability to make decisions. Thus, solutions for strengthening women's voices and agency on climate justice need to reduce women's vulnerabilities to exploitation and undue burden by increasing their equal participation and decision making in their communities, through economic opportunities, savings/ capital and safety nets that ensure they have access to land ownership, and can obtain clean energy and other technologies, family planning and health service facilities. In addition, programmes identified should start with resilient women, who can juggle a work-home balance. Environmental educational programmes should also seek to ensure that women and the youth can cope with the vagaries of climate change (Lekwauwa & Ijioma, 2012).

3.7 Other Factors

3.7.1 Financial Access

Women are disproportionately affected by poverty, that is brought about by numerous challenges. Examples of some of these challenges include: limited learning and funding opportunities. In addition, they also face social and cultural norms that hinders them from working and even if they do succeed in entering the workforce, they end up working for very little pay, for long hours and with little or no prospects for career growth (Sudha & Reshi, 2023). Women in Africa face many obstacles that prevent them from fulfilling their potential. Given multiple socio-economic pressures, women tend to engage in the informal sector (Naguib, 2022). Meanwhile, some women choose to enter the formal sector through entrepreneurship. However, Ojediran and Anderson (2020) argue that agency for women entrepreneurs, has been a key driver in increasing women's participation in business. On the other hand, a large number of women who are young, less educated and who reside within the rural populations remain excluded, from taking part in business more so in the sub-Saharan countries (Mushtaq et al. 2023). This is mostly driven by their lack of collateral and other financial securities, as well as the lack of knowledge on how to access financial capital. Besides, the supply side of financial provision, the absence of timely credit, as well as the demand side of service provision is also a hindrance to women. Women in Africa have the highest rate of business failure at 8.4 percent. However, in countries like Senegal 36.8 percent of women are entrepreneurs, this is the highest on the continent with South Africa accounting for the lowest at 5.9 percent. Ghana, Nigeria and Zambia have the highest number of female entrepreneurs compared to men (Mashapure et al., 2023). Women who are economically empowered have the ability to better contribute to their households. Mushtaq et al. (2023) cite Perrin and Weill (2022) who assert that several factors curb women from participating in the financial sector. They give examples of where women fear applying for loans due to the fear of rejection from loan officers. Secondly, women may not be able to understand the financial contracts due to lack of knowledge on financial service, and also men tend to make key decisions about household finances and the use of religion to curtail women's ability to know about financial resources (Mushtaq et al., 2023).

3.7.2 Education

Knowledge and skill acquisition can assist one to grow further within their organization and at the same time have them grow financially. Thus, it is important that one is able to have access to such training so as to give them the opportunity to start businesses and grow the economy (Sudha & Reshi, 2023). A study based on Nigerian women's education was found to play a key role in female empowerment as it was seen to increase their participation in educational policy formulation, extensive enlightenment campaigns, and removing stereotypes in the division between men's and women's jobs (Tiwari & Malati, 2023). The lack of schooling is a hindrance to women's ability to gain access to banking services (Yeyouomo et al., 2023). The importance of having such resources such as education is that it equips one with the ability to stimulate financial growth (Ojediran & Anderson, 2020). The gender gap between men and women is quite wide, and is attributed to the fact that women have low levels of education and are discriminated against because of their gender. In India, for example women are viewed as second class citizens when compared to men, due to social and gender norms.

Women are forced into early marriages as a way of supporting their families out of poverty and hence are unable to continue with their schooling. Thus, curtailing their ability to work in formal sectors and thus have better wages. However, other women have taken up the challenge to uplift their fellow women out of poverty by engaging to learn new skills and crafts that can lift their income levels (Ojediran & Anderson, 2020). Such initiatives, can be supported by government through tax breaks and can therefore encourage more women to take up businesses. In addition, the development of infrastructure such as roads and telecommunications can help business grow to new heights (Sudha & Reshi, 2023). The growing of new ventures, can thus encourage other women to guide and mentor fellow women in business and hence grow their economic capabilities with the view of increasing their income, especially in rural communities.

3.7.3 Socio-Cultural

Men are seen as working in many sectors of the business world, and it has become difficult to view women in the same light in the same sectors. This has had negative connotations on women's role as entrepreneurs (Jaiyeola & Adeyeye, 2021). In African society, women are forbidden from aspiring to greater roles, and this limits their ability to engage in different sectors of the economy, for example in Nigeria, where women are bestowed, an inferior status compared to men (Egwurube, 2016). In Nigerian society women play second fiddle when it comes to decision making, and one which is further propagated by culture and religious norms. These norms, thus strip women of their identity. Women therefore take up their partners identities (husbands). However, even after taking up the identities of their husbands, women are still not viewed as full members of the kinship of their husbands, and have to accrue benefits as such as land or property via marriage. Thus, women in this society face a lot of challenges when it comes to growing and developing their homes (Jaiyeola & Adeyeye, 2021).

Jaiyeola and Adeyeye argue that this social standing is particularly strong in the South-South and South Eastern zones in Nigeria, where such norms are conferred as unwritten laws within communities. It is for this reason that women in most African countries are unable to own and inherit land, as such privileges are set for the male members of the family, therefore women are unable to obtain additional funds from financial institutions by using property as collateral. Women's economic empowerment is thus seen as a critical tool in reducing the existing gender gaps. The building of women's ability in financial literacy is key in helping to combat poverty and improve economic outcomes for families in rural settings hence, the reason why women's economic empowerment and gender equality are part of the Sustainable Development Goals (SDGs) (Jaiyeola & Adeyeye, 2021).

3.7.4 Labour Markets

Labour market outcomes can be divided in to those that focus on (1) structural constraints and (2) individuals (Kabeer et al., 2021). This has led to the development of social norm approaches and structural constraints being added as part of theoretical frameworks. Women's work over the years has been treated as being 'inferior' because women have been viewed as being inferior, thus drawing on Phillips and Taylor's (1980) empirical evidence that suggested that definitions of skill in the workplace were often based on the identity of the person carrying out the job rather than on the technical demands of the job. Thus, women were discriminated upon based on their gender rather than their work (Kabeer et al., 2021).

This showcases the struggle that women have had to undergo over the years, with employers exploiting them just because of their sex. In short, employers applied the divided and conquer rule by discriminating against women as the weaker sex, than on men, hence using them to make profit while paying them lower wages. Gender inequalities, thus, highlight the challenges that women face when it comes to the labour market. In that, they are exploited as workers compared to their male counterparts who are seen as more able to work within extreme conditions (Grantham et al., 2021). Such misgivings, have enabled men to come collectively and organize trade unions to fight for their rights unlike women (Kabeer et al., 2021). For example, in Central Guatemala women received only 3 percent of agro processing contracts, while in Kenya less than 10 percent of women farmers received contracts for horticultural production (Dolan, 2001); while in China men received the contract farming. The reason for this is that women do not have ownership of land unlike men, hence, men are able to receive such contracts as they have rights and hence can guarantee a steady supply of produce.

To challenge such inequities requires the creation of inclusive income generating opportunities within the community (Asadullah & Kambhampati, 2021). The creation of economic opportunities for women for example in Asia through export processing zones, has accorded them the opportunity to earn money. If we therefore seek to have women as part of this circle we must endeavor to grant them access to resources and markets. They also require access to markets and networks so as to build social networks that are critical in helping them grow their businesses. In Asia, women are not allowed to go to markets, and marketing of farm produce is strictly done by men hence, the reason why women in Asia concentrate more on subsistence farming. It is for this reason, why it is important for women and girls to be educated. However, if women are not accorded the same opportunities as men, parents will not see the relevance of having to educate their daughters thus, continuing the cycle of poverty among women and girls as most of them work in their family farms without being paid (Asadullah & Kambhampati, 2021). The use of digital technologies can help increase market access for women. In addition, monetization of women's labour can help value chains and labour standards for women in agriculture (Asadullah & Kambhampati, 2021). It is therefore important that agricultural farms are able to balance women's involvement on the farm, with that of the household tasks. Furthermore, this does not mean removing men from such tasks, but rather balancing the time burden with the goal of the negotiation of power in familial and societal contexts. This can be done by incentivizing male members of the family to register businesses in the wife's name. For example, in Indonesia women were much happier making decisions jointly with their husbands than alone (Doss & Quisumbing, 2020). When farming is made a fair experience rather than a forceful one, women are more likely to take up more opportunities as they feel the responsibilities of the tasks are shared (Fernandez et al., 2015). On the other hand, we cannot assume that such opportunities will draw women to work more on the farm (Kabeer et al., 2021). This is because with added work, comes more responsibilities for women and not all work is empowering. Therefore, we cannot assume that all women would want to work. Studies conducted by the Pathways of Women's Empowerment research programme indicated that women who took up employment reported positive outcomes that were related to positive empowerment indictors compared to inactive women. The positive attributions were tied to cognitive, relational and behavioural as well as material issues.

In this formal setting employment was viewed as empowering. This was mostly generated by the public sector and less by the private firms and the NGOs. Thus, formal settings offer better job security and better wages and appeared to have less forms of gender discrimination. In Egypt for example, the state was the only genuine equal opportunity employer, however, opportunities were also shrinking as opportunities were not the same for women's access to formal private employment. The importance of work allowed women to have some form of control over their incomes and helped them be part of the public domain. However, when it came to unpaid productive work women were not motivated to participate. To have more female farmers participate in cultivation according to Asadullah and Kambhampati (2021), there is a need to change educational training to female empowerment in secondary schools so as to improve their access to natural resources such as land, via partnership with men. This is critical as most female farmers do not have secondary schooling, with this food insecurity would likely decrease. Education and training is seen as being critical in helping to equip women with life long skills as well as increase the employability prospects, which boosts their confidence levels. Gender inequalities in the labour market persist as a result of shifts to more employment that is centered on growth. Hence, the importance of understanding how such constraints affect different social economic contexts with a view of changing them (Grantham et al., 2021). Another challenge, lies with government and policy makers, and the need for them to equip women with rights within and outside their marriages especially in Asia and Africa. This will be critical with the change from subsistence farming to cash crop farming. There will be an increase in the need to have market access as well as assets ownership as has been the case for women in South Asia. Women suffer from exclusion from leadership related positions from land resolutions systems (Asadullah & Kambhampati, 2021). This can be especially true for women who come from minority communities who do have full constitutional rights. This is then further complicated by social and inheritance rights that tend to favour men over women (Asadullah & Kambhampati, 2021). Although, this is not the case everywhere. For example, in Nepal (2015–2035) Agricultural Strategy for Development for land registration, gave tax waivers to land owned by women, so as to have them gain land use rights. In Bangladesh, reforms on land ownership, have given women land tenure access (Halim, 2015). Thus, it is important that such practices start earlier rather than later in the life course (Grantham et al., 2021).

3.8 Conclusion

Climate change affects both men and women. However, as discussed, its ramifications are more severe for women than men. To effectively reduce environmental outcomes all genders must be empowered. One approach to combating climate change is the development of policies that encourage the rapid adoption of climate-smart agriculture. In particular, policies should focus on educating both men and women about new climate-smart technologies. However, these technologies and policies must be tailored to the specific communities in which they are implemented, taking into account social norms and customs. For instance, as reported by Sultana in Bangladesh, water accessibility remains a challenge influenced by societal norms. There is also a call to move beyond solely relying on Western methods of climate change mitigation. Instead, incorporating traditional practices from African communities and tribes can serve as a more affordable and effective strategy for helping rural populations adapt to climate change.

CHAPTER 4

Inclusive Agriculture: Advancing Women's Participation and Equity in the Sector

4.1 Introduction

4.1.1 Women and Food Security

In the coming years, feeding the world will become a challenge due to the increase in population, and other environmental challenges affecting food production. The challenge with food insecurity is that it is largely an income and distribution challenge. Furthermore, there has been little or no research on the impact of farming diversity in increasing food production. Research shows that agroecosystems will do better under current environmental conditions if correct measures are taken to address the present-day situations. These adaptations are already visible in Africa, where farmers are adapting to climatic and environmental changes by diversifying their agricultural production with a view of increasing their income through diverse crops and livestock production (Waha et al., 2018).

Agricultural production in sub-Saharan Africa is mostly done by women who provide 90 percent of household water supply, 70 percent of field labour and provide 60–80 percent of household meals. The irony in this is that ownership of land does not belong to them, and they are the ones affected the most when food supply dwindles. In addition, women spend a lot of their time growing food for subsistence consumption as well as engaging in care work for other household dependents. The relationship that women have with food has played a complex role in shaping

M. O. Jimoh and S. Narsiah, *Unveiling the Power of Women in Irrigation*, https://doi.org/10.1007/978-3-031-94174-0_4

their gendered identities as women. This identity is factored in social spheres (the bride price for example) and given economic significance in pre-capitalist economies. This identity is appreciated by the community and the household as women work to provide food for their families through reproductive and productive labour. Thus, the term 'Mama' is a reflection of their hard work and appreciation of what they do to sustain the community and the family. However, this honor although bestowed on women, does not in itself showcase the harsh reality of their absence in owning assets and being part of key decision-making abilities within their homes (Salo, 2023).

Food in Africa is a women issue. The reason for this is attached to three critical views: one they are major contributors to food, but still remain in the shadows despite their hard work. This is because economic systems do not attach economic value to the work women do in food production. Two, the agricultural sector has turned food production for export and not for subsistence production to meet local needs, and lastly women in the sector are seen as farm workers or just individual business women. There has been a call by the World Bank to shift this view of women and hence incorporate them as part the agricultural ecosystem. A call known as 'smart economics' where women are part of the agricultural system as waged workers or food processers. However, this view has been challenged by ecofeminists Vandana Shiva and Nancy Fraser who find the move counterproductive, as it does not tackle the systemic barriers embedded in economic, social, and political institutions that disadvantage women. This view has been critiqued as not serving gender equality goals (Calkin, 2015; Benería et al., 2016).

Furthermore, the addition of women to development programs has also been critiqued as it transposes gendered norms by simply adding more work to women while devaluing the work they undertake within their households as mothers and subsistence farmers (Farhall & Rickards, 2021). Farhall and Rickards critique is that the most important factors about empowerment are challenged and are replaced with individual capacity building. Furthermore, that power and privilege in communities hinder the key aspects of empowerment within certain aspects such as targeted and measured empowerment being cast to endpoints rather than being looked at as part of processes. That interventions taken by changing legal frameworks and other policy shifts in education and finance are measured and are seen as supporting women but in hindsight they should not be seen as change but rather as substitutes for empowerment (Cornwall & Rivas, 2015). Thirdly, that despite women's role in the agricultural sector, they are still unrecognized in many other sectors. In Ghana, women have worked as salt miners and are still unrecognized. While, the sector may not be directly involved in agriculture, it still contributes to the African economy. The complacency by corporate capital in intimidating women out of the productive economy is just another example of them removing women from income generation. These organisations, do not seem to understand the important role that women play within the food production eco-system and the sensitivity in which changes in the sector may have on food production. Women, have thus taken it upon themselves to feed themselves, their families and communities with a view of having an income and contributing to their communities despite the many challenges they face.

4.1.2 Women and Youth

In African countries young people and women contribute 24 percent and 56 percent of the labour that is used in agricultural production (Kadzamira et al., 2024). Forty percent of the agricultural labour force in sub-Saharan is made up of females, with a significant additional proportion being made up of the younger generation. However, women and young people are excluded from the process of decision making as they lack ownership of the critical productive assets. To address this issue there is a need to actively call on them as suppliers in the agricultural sector with a view of empowering them economically. The involvement of young people and women in the agricultural sector is crucial in having better hindsight of the link between young people and the sector (Kote et al., 2024). For example, young people in rural areas have limited chances of starting businesses as they lack access to skills development, finance, land as well as poor access to digital technologies. The low uptake of new technologies by farmers has thus been attributed to the low levels of inclusion of women and young people due to social-cultural constraints (Yami et al., 2024). Young people need to be engaged as active participants rather than passive ones. However, for this to happen there needs to be an open call by governments and organizations to create an environment where policies and investment can take place in order to facilitate growth in the agricultural sector. Agriprenuership has been named as the venture that can be used to tackle unemployment and draw young people towards the agricultural sector (Adeyanju et al., 2023). Adeyanju views this link as an avenue to have agricultural activities that are creative and can generate income and profits. In this, job creation is created and poverty reduced among the young people, and an increase in the labour force and in food production.

This call has been heeded by some African governments, who have seen the potential that the youth have to offer for the continent and have developed interventions that give training. For example, in Uganda and Kenya, two initiatives (Youth Inspiring Youth in Agriculture (YIYA), Kenya Youth Agribusiness Strategy, the (ENABLE) Empowering Novel Agribusiness-Led Employment) programmes are in 19 African countries. With this, for example 55 percent of young people in Uganda are employed in the agricultural sector. Thus, the agricultural sector can have a significant impact on young people as a means of livelihood and as a means to increase agricultural productivity through the value chain. Women's involvement in such avenues ensures better livelihoods for those working in the agricultural sector in the country and ensures equity across the different value chains (Adefare et al., 2024). However, women are held back by two factors, when it comes to agriculture: poor access to land and poor access to information technology (Mathis, 2024). These challenges only go on to exacerbate issues, like child malnutrition and child deaths. When women do not have access to the necessary productive resources (market information, technology, finance, extension services, land, inputs) then their participation in agriculture is hindered, and results in a continent of undernourished children and individuals, especially for those from rural settings. If these resources are provided the impact will result in increased food production and better living for children and those in the continent (Ingutia & Sumelius, 2021).

4.1.3 Women's Empowerment

Mayoux (2001) is cited by Porter et al. (2019: 181) as defining the term 'empowerment' 'as the process through which those who are currently disadvantaged achieve equal rights and resources.' Seyfi et al. (2020) view empowerment as a multidimensional social process that helps individuals to gain control over factors that affect their lives. Notions of empowerment, and women's empowerment in particular, have increasingly moved to the core of understandings of sustainability and the SDGs in particular. On the hand, the term has been abused according to Porter et al.

(2019). To him the word showcases false hope with no goal. The word is not about financial progress but rather about social change that respects women's progress, with a view of having them informed about systems that affect their lives. The 2015 Development Agenda, places food security as central to the 17 Sustainable Development Goals (SDGs). The link between sustainable agriculture, gender equality, and healthy lifestyles is supported by SDG 2. One of the many challenges that women face is their inability to own land, that they farm, and that they farm unproductive land. Secondly, they are unable to gain access to credit and other financial resources as well as poor crop varieties, making them unlikely to adopt new methods of farming (Asadullah & Kambhampati, 2021). This situation is totally different in countries like Philippines, Indonesia, Thailand and Myanmar, where there is accessibility to land and inputs regardless of gender and that women have a greater say on household income than men (Akter et al., 2017). Third that men grow commercially viable crops, while women grow subsistence crops. This action in itself, has predisposed them to poverty due to lack of resources to farm.

The lack empowerment female farmers, thus forces them to take up work as temporary workers, or in micro business as sellers. These initiatives thus, do not help them develop to their full potential and goes on to increase the gender gap within the agricultural households. One of the major factors hindering development is the unavailability of sufficient food (Chege et al., 2021). Efforts have been made to tackle this challenge, but as many as 815 million people globally are under nourished. Those who suffer the most come from sub-Saharan Africa. Diet changes have been identified as being the best solution to fight malnutrition (Chege et al., 2021). Women play a critical role when it comes to improvising diets within their households, as they are the main producers of food. This is especially true for those residing within rural communities as, they often grow what they eat. Thus, growing more diverse crops has the potential of reducing malnutrition. However, because of their inability to access productive resources their participation in farm activities is threatened, and thus reduces their potential to increase food production and better diets. The way to increase agricultural productivity is to close the gender gap. However, gender inequalities between men and women continue to widen the development outcomes (Njuki et al., 2016). Women from rural backgrounds face a barrage of challenges that hinder their ability to farm and poses huge social and economic costs on the community. It is argued that by empowering them, with the right productive resources they can increase their income and spend more on their food, health and children's education more than men do. The importance of having women as workers in agriculture, and especially in agricultural supply chains is critical for them to have access to food and as means of livelihood.

In West Africa 43 percent of the population continues to work in activities in agriculture or linked to the sector, with a good proportion having a significant portion of the workforce. Nineteen-point one six percent (19.16) of the total employment in the formal wage force is dominated by men. With such figures and an increasing population, unemployment is bound to be high for young people and female workers. Seventy percent of all workers and 80 percent of female workers in Senegal are employed in insecure jobs that are part of informal economy (Buvinic et al., 2020). Furthermore, most of the working opportunities in 8 out of the 17 countries in West Africa are restricted to women and marginalized groups, with men occupying a majority of the key roles. These countries have numerous natural resources, but have created limited job opportunities outside those linked to natural resources (Buvinic et al., 2020). To change this narrative it is important that countries initiate programmes that can change agrarian economies, with the motivation to empower women and increase their agricultural productivity and change their economic realities. Programmes that seek to empower women have placed childcare, entrepreneurship training, and financial training and grants at the centre of improving women's skills in deprived areas (Ambler et al., 2021). Those, that have been successful according to recent studies, focused on core challenges that women faced; childcare, improved financial status and the availability of job services. Additional interventions have been the improvement of electricity provision in rural settings (Buvinic & O'Donnell, 2016). The impact of these interventions is measured within households, with the disaggregated data being limited to women headed households. However, this is contrary to what is present on the ground, as most women are married and live in male headed households, thus, female headed households are not a valid representation of households in general.

This is significant as women's empowerment within married households is dependent on their partner, when making decisions. Thus, it is important to ensure partners of the women are receptive of empowerment efforts. The Feed the Future program developed the Women's Empowerment in Agriculture Index (WEAI) (Alkire et al., 2012, 2013), which has been able to provide means that can measure women's progress in agricultural programmes. This measurement system has had a great impact on the agricultural sector as it seeks to improve knowledge about women and their connections to food security, empowerment and agricultural growth. Furthermore, it helps to identify obstacles, with a view of allowing women to identify disempowerment and provide ways in which they can increase their agency in critical areas. This tool factors five critical areas: (1) time management, (2) executive management, (3) financial control (4) access to, and authority over, key livelihood resources, (5) production choices in agriculture (Njuki et al., 2016). It is important to note that job creation outside the home will improve women's collective agency. For agricultural production to improve there is a need to include women and empower them significantly within the agricultural sector, with a view of increasing food security (Asadullah & Kambhampati, 2021). Women's participation in food production is linked to improved diets and health for themselves and their families. The participation of men and women in food production is acknowledged. However, their engagement is dependent on the structure of the economy and gender norms. This is because care work responsibilities for women, has a direct impact on their work within and outside the home which has a direct impact on their diets. It is therefore important that women's inclusion in the sector be seen as a way to empower women as they make critical life decisions (Quisumbing et al., 2021). The new and varied changes taking place within production value chains and food systems has the ability to offer new opportunities for women and also the opportunity to create new barriers as participation does not immediately transcend to benefits, especially if the terms do not favour women. For instance, as more people move to urban centres, more women will increasingly work away from home. This will then result, in them relying more on the open market than on their own subsistence production, which will in turn affect their nutritional diets and health. Current diets include super-processed foods, that are high in sugar, fat and sodium. Furthermore, these foods are seen as leading to chronic diseases. However, home styled processed foods that are common and available in local value chains may retain nutritional value while reducing women's load (Quisumbing et al., 2021).

4.1.4 *Investing in Women in Agriculture*

Numerous papers have documented the importance of empowering women and the impact of empowerment on women and society. Women's empowerment has increased benefits on child investment and poverty reduction. However, little is known about policies and programmes that can increase economic opportunities for them and the next generation (Buvinic et al., 2020). Within the agricultural sector women are viewed as key players, as they are involved in both the production and marketing of products (Jimbira & Hathie, 2020). Women comprise of 52 percent of the total population who work in this sector in sub–Saharan Africa and assist in 60–80 percent of food production in Africa. In addition, they comprise 50 percent of the agricultural labour force in the sector. There is a strong call for women to capitalize on the agricultural sector in Africa.

However, for this to happen, women need to have access to land and finance. The African Human Development Report of 2016 estimates that 95 USD billion per year is lost in the continent due to gender inequality. Another document by the McKinsey Global Institute in 2015 notes that 12–28 USD trillion can be added to the world economy if women have equal opportunities as men (Anderson et al., 2020). Entrepreneurship is viewed as a key driving force in helping to curb poverty in the Global South, as it opens up avenues for job creation and wealth, while building confidence in individuals. It is for this reason that NGOs and governments have embraced entrepreneurship (Ojediran & Anderson, 2020). On the other hand, this focus should be on women, with the view of building their skills and involving them in niche markets, as increasing their farm output alone will never be enough (Jimbira & Hathie, 2020). The call for women to take up business opportunities is seen as a window for them to build financial independence and to have a say in their activities, independent from male members of their family (Ojediran & Anderson, 2020). Alkhaled and Berglund (2018: 877) summarize this as 'women are good for entrepreneurship and that entrepreneurship is good for women.' As powerful as the statement sounds, it neglects to consider the barriers and challenges that women face with regard to the work they do (Ojediran & Anderson, 2020). For instance, women's involvement in agricultural research is a good example of closing the gender gap as it will allow for growth and development of the agricultural sector in Africa, through the development of gender sensitive programs and innovative solutions. Current figures assert a steady increase in the number of women in agricultural research as rising to 15,000 in 2014 from 9000 in 2000 an increase of 24 percent (Jimbira & Hathie, 2020).

Investment Challenges for Women in Agriculture
The involvement of women in the agricultural sector is hampered by numerous barriers and challenges. For one, they have limited time to focus on farming as they are also engaged in other household tasks, that are crucial for their household management. They also grow crops for subsistence use and not for commercial reasons, hence they are disadvantaged in taking up opportunities that can grow them economically (Anderson et al., 2020). Female farmers tend to have 20–30 percent lower yields than men in sub–Saharan Africa (Jimbira & Hathie, 2020). Hence, empowering them economically has numerous gains as this can be translated in a change of social and mental mindsets, which has the ability to improve their agency (Anderson et al., 2020). Anderson et al. (2020) challenge the Women's Empowerment in Agriculture Index (WEAI) as not testing diverse economic returns by gender for a specific intervention, and that few studies have been done to evaluate the economic returns of interventions targeting women's economic empowerment in agriculture. Increasing women's access to land, labour and capital is seen as the first step in improving women's empowerment in agriculture and harnessing the differences between males and females.

The argument around increasing access is that when both genders have access to the same resources and opportunities, they will be equal in their outputs in that women start at a lower level than men when gaining access to resources and that their use of inputs will increase when given more access as compared to investing in men. The second argument is that men and women's choices with regard to household choices is different and that by increasing women's agency, opportunities open up for women to be part of the decision making within the household, and therefore affect economic outcomes within their homes (Anderson et al., 2020).

Land—The Scarce Asset
Land is a critical asset as it revolves around people's livelihoods and ensures household food sufficiency and cultural identity. The cause of hunger and poverty has been attributed to the scarcity of this resource. Worldwide, one in out of five small holder farmers suffer from hunger and do not own this critical resource (Fitawek et al., 2020). The ability

for women to own this scarce resource is seen an avenue for empowering women and enabling them to promote economic growth and development (Doss & Meinzen-Dick, 2020).

According to Doss and Meinzen-Dick (2020) women's ownership of land is crucial in meeting their financial needs. This they state has been found as being critical in both academia, policy and published work. Evidence has been provided by Budlender and Alma (2011) on the positive outcome of secure land rights. Additional evidence by Meinzen-Dick et al. (2019) draws on the theory developed through the Gender, Agriculture and Assets Project as reducing poverty among women. This has been attributed to the following: reduced gender-based violence, agency, entrepreneurship, better food accessibility and intergenerational transfers (Doss & Meinzen-Dick, 2020).

Several indices have been linked to the increase of agricultural outputs globally since the mid-1990s. The use of better crop varieties, fertilizers and mechanization have been attributed to this (Gashu et al., 2019). However, Africa, has 52 percent of the worlds arable land, and has a high population growth rate. This increase is threatening agricultural land for production as there is increased demand for industry, roads and other infrastructure. The loss of this land is estimated to reach 30-60 million hectares by the year 2030 (Gashu et al., 2019).

Factors Influencing the Evolution of Farm Structures in Africa

Land sizes across the continent are evolving. These changes have been linked to: high food prices, increase in the value of land, change in agricultural policies and lobbying (Jayne et al., 2019).

The Rise of Land Markets

Land acquisition for the next generation is hanging on a whim. This has been attributed to changes in customary land rights which are increasingly becoming privatized. This change is being linked to a rise in private farmers, for example in countries like Zambia and Malawi. Furthermore, changes in land policy have made these changes possible, with rural people and migrants now, not having the possibility of owning land via inheritance due to the increasing demand by wealthy outsiders for the vast open lands. These actions have increased the value of the land and have transferred customary land tenure to move to statutory tenures that have long term tenancy rights (Jayne et al., 2019).

High Global Food Prices

Food production has declined in the last few years and led to an increase in food insecurity. This has been linked to diseases that have affected animals and other animal products (Larbode et al., 2020). Glauber et al. (2023) notes that the COVID 19 pandemic has also been linked to the higher food prices across the globe. In the past, trade policies were used to manage food supply issues. The Russia–Ukraine war highlights this example. The two countries alone account for 12 percent of calories traded (grains, and vegetable oils). The war and the COVID 19 pandemic have led to an increase in prices of grain, with prices ranging in figures last seen in 2007/08 and 2010/11. In addition, export restrictions, have further led to other countries initiating measures that reduced the availability of the commodities and caused a hike in prices globally, thus increasing the supply and demand challenges and causing panic among consumers.

Policy Reforms

The changes in agricultural market policies taken in the 1980s and 1990s were crucial in eliminating barriers to private trade. Furthermore, the impact of these changes, then became visible in the mid 2000s, when the prices of food increased thus, making private organizations venture in to profitable initiatives that led to the growth of regional agri-food systems. The reduction of price control mechanisms also led to the better pricing of local produce in relation to that which was being sold on the global markets. These changes have therefore led to more private organizations investing in agricultural value chains that are now transforming the region and the sector (Jayne et al., 2019).

Farm lobbies and political capture

Between 1985–1995 agricultural policy struggles among African governments and international lenders, resulted in local interests taking over the policy agenda in the early 2000s. These changes led to the increased voices of national farmer unions who fought for their interests. However, because farmers do not belong to the same unions or groups, policies affect them differently depending on which union or group they belong. For example, some farmer unions may call for agricultural subsidies and land allocation that ends up benefiting large farms instead of small-scale farmers. Farmer unions are largely seen as backing policies that favour increased prices of food that support large farm producers, changes in customary land to private land, and offers of credit subsidies to bigger farms. The reason for this is that they want farming to be seen as a business, that can transform African farming. These views represent the new generation of farmers who have been subsumed under a capitalist system (Sitko & Jayne, 2014). This generation is enlightened and has excellent connections with those in authority, hence are able to take advantage of land polices and institutions which in turn benefit their own interests. Due, to the increase in global food costs the demand for market oriented farming has gone up by driving the need to purchase land within the continent. The transformation of the agricultural sector in Africa is possible, only if African governments agree to support small scale farmers with incentives and policies that can increase their productivity. Countries such as Rwanda and Ethiopia appear to be engaging with this model. However, the increasing number of commercial farmers will influence the manner in which agricultural support is given to those owning large farms around the world (Jayne et al., 2019). This is because LALIs (large-scale agricultural land investments) have affected environmental eco-systems and caused an increase in conflicts, and poor employment opportunities as they offer short term labour. In addition, the support offered to women is lacking in terms of supporting them with regard to socio-economic shocks (Osabuohien et al., 2019). By having better access to land, labour, and social engagement, plantations will increase production. This has gender implications. The reason for this is that women lose their current land rights, when their small farms are acquired, hence affecting their ability to produce food and earn income. Thus, the bottom line of these changes lies in the power relations that allow or deter access (Fonjong & Gyapong, 2021). This, then has an impact on the eco-system. This development could then have a great impact on the model used, the type of crop grown, and the kind of investment made especially on lands that are not in use. In addition, how do large farms adapt to existing customary tenure systems. This is because these traditional systems offer women usufruct rights, that allows them to grow crops for subsistence use, and thus encourages gender disparity with regard to access and control over the land. In rural West Africa, for example, resources are not organized based on norms, or sharing, hence, distribution of assets such as land is not based on equity. When such nuances are added on to current land markets, they only seek to increase the challenges women face with regard to accessibility, thus, challenging how land accessibility is given and granted. Drawing on the need for checking on gendered relations in large land transactions. The need to include women's work in household economic production has been called upon by feminist agrarian political economists, following Marx (1977). This is because their accessibility to land as small scale farmers is linked to their relationship to men who control the land and other resources. Women tend to lose this access when their land is consolidated into large spheres, thereby rendering them landless and without means to fend for themselves. Fonjong and Gyapong (2021) draw on the importance of women's roles in social reproduction as land access and labour relations are at the center of this. Food production is critical as large land consolidations can reduce women's abilities to produce food for their families and their communities.

The example of smallholder farmers in Ghana, Nigeria, and Mozambique highlights a negative correlation between tenure security and the level of local land market activity (Jayne et al., 2019). Research suggests that households in areas where land is of high value, face an increased chance of land loss due to personal and state acquisitions. Notably, traditional land holding arrangements face increasing pressure as land becomes more commodified. The traditional protections that once safeguarded individual land rights are gradually weakening, leaving smallholders more vulnerable to land insecurity and displacement.

Historically, one key difference between economically disadvantaged rural families in sub-Saharan Africa and Asia countries was that, in Africa, a good number of households had at least some rights to land (Badiane, 2019). However, changes in land use have disproportionately affected women, particularly in terms of food security (Fonjong & Gyapong, 2021). Women rely more heavily on land for subsistence farming, whereas men are primarily engaged in cash crop production. For instance, the leader of the Women's Forum Ndian voiced her frustration over land enclosures affecting local communities: 'The women are not happy at all ... our land has been taken ... we no longer have farmland to grow food to feed our children ... women now travel long distances to cultivate in areas that do not have roads ... Food is scarce today, and the women are anxious.' In Pendamboko, where the Cameroon Development Corporation (CDC) is increasing farms, many female farmers have been forced to plant on low productive land. Women in these communities take pride in their ability to farm independently, no matter how small the plots, often intercropping alongside family farms (Fonjong & Gyapong, 2021; see also Amanor-Wilks, 2009). Men take charge of high value crops such as cocoa;

women cultivate subsistence crops like vegetables, and legumes. These plants are important for home consumption and income generation.

The increased number of farms has worsened food insecurity, particularly for migrant female farmers who come from households and indigenous groups who do have secure land rights. For example, with complete dispossession, women must now acquire land from clan members under sharecropping agreements a system traditionally reserved for temporary settlers but now applied even to displaced indigenous communities. This arrangement severely limits the types of crops they can grow, the scale of their production, and their share of the harvest. As a result, many women with restricted land access face increasing barriers with regard to engaging in independent crop production. This has resulted in them planting diverse crops, while others have resorted to look for farming land that is further from their homes and have to resort to shared crop farming, which affects the amount of family labour they can receive from the extended family. This in turn affects their output. These are conveyed by a woman affected by this, who states: 'I used to farm vegetables, corn, groundnuts, maize and my husband also cultivated these …in addition to maize, rice and plantains. I usually marketed my husband's produce for him, and I keep my own income. After the takeover, I now sharecrop 1 acre of rice, groundnut and corn and this land is distant from my husband's sharecrop land' (Fonjong & Gyapong, 2021: 6). Land tenure is complicated by the varying nature of the interventions and institutional arrangements. It is clear that land policies in Africa need to recognize the continued existence of customary land rights and customary land administration institutions if there is to be sustainable impact on perceived tenure security and agricultural development.

Where customary land institutions remain relevant, as in most rural contexts in Africa, land policies should either formally recognize the authorities of these institutions (subject to conformity with national policy principles and objectives). For instance, strengthening women's rights by incorporating them into more comprehensive formal land administration institutions where they are most likely to participate. For example, Zambia recognizes the authority of customary institutions to manage customary rural land rights, and a number of other land policies in the region have proscribed legal roles for customary leaders in land registration and administration (e.g., Botswana, Côte d'Ivoire, Ghana, Kenya) (Jayne et al., 2019). These policies represent a more 'fit for purpose' approach to land administration that seeks to address current land tenure realities rather than attempting to replace existing customary institutions or introduce onerous technical standards that go beyond what is currently required to secure land rights. Given the diverse nature of customary land tenure systems in Africa and threats to tenure security facing smallholder farmers, policies to strengthen tenure security and regulate land transactions in Africa will need to be carefully tailored to the local tenure context and needs of different landholders to affect perceived tenure security and agricultural outcomes. Regardless of the institutional form (customary/government/hybrid) and level of land registration (individual/community), experience suggests that land administration systems must adhere to a number of other principles to positively impact tenure security and rural transformation (Jayne et al., 2019). To be effective in increasing perceived tenure security, land administration systems must be seen as legitimate by all stakeholders, including the most vulnerable landholders, as well as large private investors (Jayne et al., 2019). Second, to be sustainable (both fiscally and in terms of their ability to reflect land transactions), land administration must be decentralized to a level that balances user accessibility (and demand for land services) with recurrent administrative costs. This remains an elusive objective for most land administration systems in Africa and calls into question the long-term sustainability of recent land registration interventions. Finally, it is increasingly acknowledged that secure land rights are a necessary but insufficient condition to promote on-farm investment, productivity growth, broader multiplier/growth linkages between on-farm and off-farm development, and rural economic transformation. To accomplish these broader policy objectives, land administration systems must be linked to complementary institutions and services. These include customary and formal dispute resolution mechanisms, as well as rural finance, utility services, and agricultural value chains, for example by linking service provision to recognized land rights and ensuring that land registration beneficiaries have access to agricultural input and output markets (Jayne et al., 2019).

Water Scarcity

The interdependencies that exist between water, energy, land and food are numerous and multidimensional, and these interlinkages are referred to as the water-energy-land-food (WELF) nexus. These resources are crucial to human existence and how they are governed affects outcomes in terms of social equity, externalities, and socio-ecological resilience. Moreover, sector policies for these resources are intertwined, in particular, in their trade-offs. Policies for one sector often bear consequences or negative externalities for the other three sectors, on local, national, regional or global scales.

These interlinkages add to current pressures on water and land as well as on resources that fuel the energy system, exacerbating existing scarcity problems, as the demand for food, water and energy is expected to increase by 30–40 percent by 2030 (German Development Institute, 2013). Therefore, the implementation of the nexus requires policies, institutional arrangements and procedures that can take these trade-offs and synergies into account (German Development Institute, 2013). Failure to do so will generate high costs to the economies of SSA and exacerbate inequalities across countries and social groups, both now and in the future (Hlahla, 2022).

Women, who constitute two-thirds of the world's poor, are key stakeholders in nexus sectors, however, they face significant structural barriers. Moreover, the working culture in these sectors is still guided or influenced by masculine attitudes, behaviors, and mindsets and women struggle to access these men-dominated spaces. Very little effort is made to understand and tackle systemic gender issues related to water, energy, land, and food within the nexus sectors (Hlahla, 2022). The effects of climate change on nature and people are mostly experienced in developing countries where a majority of the rural poor reside (Tantoh et al., 2021). In spite of this, the WLF (Water, Land and Food Security) nexus notion continues to be examined top-down with women mostly disenfranchised in the access and management of water and land in particular, coupled with isolated sectoral approaches of policy intervention developments. Likewise, risks linked with climate crisis appear to consolidate gender disparities. Gender issues and climate change are cross-cutting priorities for many national governments. The importance of a gender approach and gender mainstreaming in natural resource management has been identified as one of the most contemporary ecological problems in the developing world (Tantoh et al., 2021). Ingutia and Sumelius (2022) assert that water, food security, nutrition, health, and women's empowerment are all closely interlinked. A conducive environment that facilitates women farmers' adoption of irrigation water technologies would increase their chances of having reliable access to water, subsequently leading to several positive outcomes, including intensified cultivation. This in turn would lead to increased productivity throughout the year and greater availability and stability of food supplies during the dry season, enabling crop diversification in micronutrient-rich vegetables and fruits and thereby enhancing the nutrition status of crops. The availability of irrigation water mitigates drought, resulting in improved crop yields and crop quality and potentially promoting food security.

Access to water and its distribution is influenced greatly by gender due to the socio-cultural, political, and economic roles assigned to men and women. In rural communities in most developing countries, the role of sourcing water for household use usually falls on women and girls. This increases the exposure of women and girls to a high risk of assaults and physical trauma. Emerging evidence indicates that violence experienced during water acquisition in times of shortages is largely gendered and disadvantages women and girls more. Violence experienced at the water point comes in the form of physical violence, psychological violence, and sexual violence. Physical violence is perpetrated by both male and female adversaries and, according to feminist theory, disadvantages women and girls more due to sociocultural power structures (Chigusiwa et al., 2022).

Evidence shows that women face greater resource and agency-related constraints in adopting irrigation practices and technologies, such as less access to land and water for irrigation, less access to financial capital, restrictive social norms, lack of access to knowledge and training, and heavier work burden at home. These constraints not only limit women's ability to adopt irrigation practices and technologies but also limit their ability to benefit from them. For example, women have less influence over decisions related to the use of irrigated crops or the spending of income from the sale of irrigated crops. At the same time, irrigation interventions may also contribute to women's agency or disempowerment through changes in their control over agricultural production decisions, income decisions, and time use (Bryan & Garner, 2022).

Irrigation activities targeted towards women, for example on plots managed by women or on irrigated home gardens, have been shown to increase women's control over irrigated produce and income, and improve nutritional outcomes. Irrigation can also affect women's time in different ways. It can either relieve women's time burden or add to it depending on the type of irrigation technology being applied (e.g., either manual or motor pump). Time allocation may also shift among different family members when technologies are adopted which can influence time spent caring for children or engaging in income-generating activities. Finally, irrigation may lead to well-being outcomes (achievements) for women through several pathways and changes when women's agency pathways intersect. For example, women's involvement in agricultural and irrigation decisions has implications for production choices. These include the types of crops that are planted, how these crops are used (e.g., sold in the market or consumed at home), and how to spend the income earned from selling irrigated crops. Women's involvement also has potential positive implications for nutrition, health, and education. The resulting relationship between irrigation and women's empowerment is heavily dependent on the broader social, political, and institutional context that governs men's and women's behaviour and interactions. For example, social norms governing men's and women's roles in the household and community might prohibit women from engaging in certain activities, like irrigation using manual pumps. Women also have different preferences for irrigation practices and technologies given their socially-determined roles in the household, community, and agricultural activities, for example in Ghana even when women have access to land and water for irrigation, they lack complementary inputs, such as fertilizer or fencing (Bryan & Garner, 2022). Women irrigators reported that their lack of access to fertilizer or a delay in the timing of fertilizer application affects the productivity of their irrigated plots, which limits their ability to benefit from small-scale irrigation. Fencing is considered important because it is needed to protect irrigated plots from destruction by livestock during the dry season when they graze freely. As with the hand-dug well, fencing is built by hand each dry season using mud and sticks. Both men and women reported that women do not have the "strength" to build their own fences and do not have the financial resources to hire labour to construct fencing. Because it is a time-consuming and arduous task, men prioritize building fences around their own dry season plots and often do not invest in fencing around their wives' plots. In many countries in sub–Saharan Africa, the interconnection, between water, land and food is often ignored. However, there has been an increase in advancing the nexus and in understanding the impact of environmental adaptations on the climate crises. Water is the most consumed resource from the nexus, with 70 percent accounting for the withdrawals, furthermore, it is critical to agricultural production as it is used to produce energy and for transportation. Moreover, 30 percent of the total global energy is used for food production (Tantoh et al., 2021).

4.2 Conclusion

The path to agricultural transformation in Africa, needs to ensure that agricultural production and food availability for consumption and the market for economic development is inclusive of solutions that include women (Mapedza, 2024). The absence of changed social norms is critical in increasing agricultural productivity, as women cannot be successful transformation agents if only technical solutions are used. The need for improved agri-systems after the pandemic should be based on research that seeks to transform the agricultural eco-system, through resilience. In addition, this transformation, can be addressed when access to land and water is granted, especially for women by challenging the patriarchal notion that a farmer is man. In Malawi for instance, women agricultural advisors had to be trained on the importance of gendered social dynamics (Mapedza, 2017).

A transformative gender strategy or (GTA) Gender Transformative Approach can assist in building the intersectionality approach, with a view to help women who face multiple challenges in the agricultural sector by overcoming the barriers, which refer to access of resources and other intersectionality markers like age, race and disability. The second, is by addressing the patriarchal privileges and castes, and lastly the environmental challenges as well as technology and language (IWMI, 2020). To address this further, there is a need to ensure the availability of disaggregated data, that will assist in monitoring changes and impact, with the view of having clear knowledge with regard to the interactions on land and water rights and agency within households. This is critical in understanding the impact of social relationships and their interactions within kinship, marriage, culture, economy, and politics (Peters, 2018).

This seeks to serve women farmers better with a view to transforming the system, and thus showcases the critical need of making changes in culture while including gender in the sectors transformation. In addition, it removes the burden that has been put on women to change themselves with a view to satisfying the patriarchal food system (Quisumbing et al., 2019). Thus, for change to take place in African farming, policies must be socially inclusive, with the view of drawing up the inequality fault lines that became visible during the COVID 19 pandemic. In addition, to change the norms that sideline women and increase gender inequalities. It is by addressing these social inequalities, the true transformation can take place (Mapedza, 2024).

CHAPTER 5

Agriculture, Policy and Gender

5.1 Introduction

Africa is a continent with a large population and an abundance of natural resources. The availability of these resources serves as a strong ingredient for economic growth. However, the continent struggles with a vast array of challenges that set back development. Agriculture is seen as a key sector that can catapult the continent to overcome these challenges. Strong agricultural policies that are focussed on growing the agricultural sector are seen as critical in helping to make the necessary changes. These polices if actioned, can increase the job opportunities for the young population and ensure environmental and ecological sustainability that can sustain the sector. Thus, the need to have strong agricultural policies is critical to make these changes happen (Petr et al., 2018). The policy narrative within the years in Africa, has focused on upgrading farming from subsistence production to large scale production. This focus has been built on using modern technology and farming methods as well as consolidation of farm lands. In this large transformation commercial farming is seen as generating increased revenue through exports (Mdee et al., 2020).

The challenge with this narrative is that it stands on the backdrop of the World War discourse on industrialization of agriculture, which is far different from current needs. In addition, it is important to note that changes in policy do not necessarily change the challenges that face small scale farmers. Moreover, the need to use industrial farming methods

M. O. Jimoh and S. Narsiah, *Unveiling the Power of Women in Irrigation*, https://doi.org/10.1007/978-3-031-94174-0_5

threatens the environment that is critical food production. The drive by donor agencies to adopt 'the green way', is yet to build consensus on the way forward for sustainable agricultural farming practices and policies (Mdee et al., 2020). As noted by Mdee (2020), the lack of consensus, has brought more confusion on the way forward with regard to sustainable farming. The argument by critics is that highly connected organizations, technology enthusiasts offer the 'green way' as the better option to farming, while those grounded on the opposite side call for greater reforms in the world food chain with support for farmers and the natural environment.

Mdee et al. (2020) asserts that there is no guarantee that new farming methods will have the ability to deliver improved yields and economic transformation. What is clear is that there is strong evidence of the factors that challenge agricultural policies. For example, Chapoto et al. (2015) cited by Mdee et al. (2020) observed how increased public spending on agricultural subsidies took place at the same time when general elections were taking place. In this lies the policy rhetoric where rural farmers who are the backbone of farming and are critical voters during elections are controlled and exploited. The use of tax and price control is thus a common means by the rich to oversee the agricultural sector in pre- and post-independence eras. Policies on agriculture are thus concentrated on the ruling class at the expense of the small scale producers (Poulton, 2012). State aid given to farmers is a means to sway rural voters during elections and distract from the absence of development within their communities during and after elections. The State focuses on keeping food prices at a minimum with the objective of reducing political noise by urban dwellers who are politically active and by showcasing affordable food prices for the rural poor.

Government support for crop production is given often to the ruling class, to acquire their portions and reduce the impact of sustainable agricultural production by reducing the number of diverse crops planted, by promoting the use of modern fertilizers and thus ignoring the environmental impact of long-term use (Mdee et al., 2020). The need to have policies that focus on the inclusion of gender in development, is being acknowledged by governments, and practioners. The reason for this call has been attributed to the absence of visible impact in development projects (Kristjanson et al., 2017). Research has shown the importance of including gender as part of agricultural policies. Furthermore, it is critical to understand how gender navigates enablers and barriers that improve agricultural crop productivity and livelihoods. Women continue to make important contributions to development, despite the numerous challenges they encounter (Phiri et al., 2022). They are key supporters of environmental conservation, food production, health and household welfare which are key areas for policy makers to note (Habtezion, 2016). Food production in Africa revolves around women, due to the changes in crop production. Current research in agriculture is focused on highlighting the important role that is played by women within the African agricultural system. Thus, promoting female farmers access to agricultural resources, is viewed as one of the ways to boost crop production (Gella & Tadele, 2014).

5.2 Agricultural Policies and Gender

Gender mainstreaming is critical in policy development and is a push towards gender equity as it supports the empowerment of women. It is also critical in opening doors for them to access resources as well as improve their health and household wellbeing. This strategy was called upon during the Bejing platform of 1995, that called for States to actively implement and promote gender with a view of empowering women globally (Ampaire et al., 2019). Policy development and monitoring and evaluation (M and E) view gender integration as critical in meeting development goals (Howland et al., 2019).

Devkota et al. (2022) defines gender mainstreaming as an approach that includes minimizing gender disparities by transforming institutions and giving women the ability to be change agents, hence decreasing their vulnerability in communities. Including gender mainstreaming in polices focused on agriculture is important if change is to be visible in agricultural communities. Women are viewed as second class citizens, when compared to men, hence they face difficulties in improving their economic positions. For instance, there has been improvement in the number of women in the work force, however, little change has been visible in their representation in politics and in tackling gender based violence (Gottlieb et al., 2016).

The presence of economic empowerment though powerful, does not necessarily translate to improved incomes, as social norms continue to view women as weak. Thus, to comprehend, the gap between State uptake of world agreements and the impact on gender imbalance programmes in projects, it is important to include theories that relate to global norms. The value of this is that policy development is 'intertwined with power structures' and thus understanding this helps to build better policies that are implementable in programmes (Howland et al., 2019).

In Uganda, for example a theory used to grasp diverse translations of gender responsive frameworks in farming and climatic shift policies explained how norm negotiation and adaptation were built and changed to meet the specific needs of the diverse normative situations. On the contrary, the use of international norms, may become barriers as they may not meet up within local contexts. Thus, there is need to develop gender norms that can be easily integrated with other local norms without being taken negatively within local contexts. It is important that this process is undertaken as it is critical in helping to pick out changes in translation, periodization and interpretations that can bring about the lack of acceptance. This is because the adaptation of international policies in local contexts may not be implemented immediately and practiced (Howland et al., 2019).

Despite, increased calls for the inclusion of gender mainstreaming in agricultural policies and programmes, there remains a strong challenge with regard to its inclusion. There is a need to address the inclusion of gender, if there is to be visible change in the agricultural sector (Howland et al., 2019). The strong opposition to challenging gender norms is seen to arise from the gender that benefits from social and economic power as they have been part of culture for a long time hence, these notions are integrated without thought. The best way to combat this is by asking why they are in place and looking for better ways to change or remove them. There have been several arguments that seek to explain the gap in gender policy implementation. For example, climate change policies ignore the impact of gender in the use of technology. Women face difficulty in using new technology as they may not be well educated or well trained to use the tools due to the lack of time. Furthermore, policies that seek to place market based solutions first can disadvantage women because of their inability to access capital and collateral (Howland et al., 2019). Governments have a critical role in changing policy narratives from 'gender blind' to 'gender aware', by ensuring laws, policies and projects carry this narrative (Muralidharan et al., 2014). Muralidharan refers to this type of change as relating to the level of integration which can be 'gender-accommodating' or 'gender-transformative'. With the first approach, programmes seek to factor and work around unequal gender norms and associations without demanding current norms and gender power dynamics. The transformative approach (GTA) seeks to help societies understand the barriers social norms create between men and women, despite improving resource accessibility. Policy development is complex, and needs those involved in its development to be aware of the critical decision makers who look at their needs and preferences with a view of ensuring no negative outcomes arise. Thus, is important these observations be done at local and national levels (Howland et al., 2019).

5.3 Agricultural Policies in Africa

5.3.1 Kenya

The Kenyan constitution in Article 43(c) under the Bill of Rights asserts 'every person has the right to be free from hunger and to have adequate food of acceptable quality'. Article 60 (1) (c) calls for ecofriendly care of resources such as land while Article 69 (1) (b) reinforces the need to maintain and have 10 percent of tree cover over Kenyan borders. These statements in the 2021 agricultural policy affirm the Kenyan government commitment to protecting natural resources for current and future generations. To ensure, these actions are implemented at the grass roots level, the fourth schedule in the constitution instructs County governments to act on the same with the Kenyan state retaining the executive capabilities of policy making under Part 1 Section 29 of the Fourth Schedule. The agricultural sector is critical to Kenya's economic development, hence the incorporation of these calls for a guided development of the sector at both National and County governments (Government of Kenya, 2021).

This drive stems from Kenya's 2021 agricultural policy that is built on the constitution and the Vision 2030 policy. The document calls for the management of the sector with a view for all government agencies to work together with the private sector and other organizations to ensure the sectors growth and the management of the natural resources critical for agriculture. This ensures that the eco-system (soil, water, land, agroforestry and fishery) are well managed within the constitutional grounds and that of the Vision 2030 policy. Gender inequities are drawn upon in the document as hampering agricultural growth and the reduction of poverty, which then impacts labour outputs with regard to accessing and controlling productive resources. The National and Country government affirm this challenge as heavily impacting the agricultural sector, and note that focusing on certain aspects of the productive processes will not solve the issues at hand. They acknowledge that the main issues to be addressed are related to the norms that exacerbate gender inequities in fisheries, livestock and crop value chains. They focus on three main objectives that the National and County governments should address: 1. Include gender mainstreaming in the agricultural sector and have accessibility to productive resources and markets increase. 2. Reduce gender inequities in accessing agricultural financial support 3. Ensure new technologies are created and easily shared among communities. Important to note is that the 2021 agricultural policy does not mention the development of key sectoral policies that would be focused on addressing sectoral issues in the sector.

5.3.2 *Kenya's Gender Draft Policy*

Kenya's Agriculture Gender Policy was developed in 2012. The policy was authored by Waithanji (2021), before the devolution of power to county governments. The 2021 gender policy addressed the functions of the agricultural sector and was adjusted to meet with the current development goals. In view of this key developmental conventions were factored in its development, these include: (CEDAW) the Convention on the Elimination of all Forms of Discrimination Against Women; The Comprehensive African Agricultural Development Program (CAADP), which addresses the need for income increase in agricultural based economies in Africa and the betterment of food security, The Paris climate action that calls for the need for women to access climate smart technologies, while ensuring environments can support change in gender relations and empowerment so as to improve women's access to assets and decision making power within their homes. The main objective of the policy is "Gender equality in Kenya's agricultural sector for enhanced and equitable productivity, food security, growth and national development". Within the policy is a call for collaboration between the two levels of government (National and Regional). However, the challenge of this partnership arises from the absence of baseline disaggregated data, to measure against and that is sector specific (Waithanji, 2021). Current data is based on samples or case studies with no regional or national data available. To tackle this challenge requires one to be engaged in a systematic study that will highlight the necessary data sought. In addition, this should be done while incorporating concrete reports that include detailed monitoring and evaluation systems. This should be part of the Strategic Analysis Knowledge

Support System (SAKSS). This document is recognized by the constitution of Kenya and gives guidance on the need to have non discriminating laws that promote equity in asset ownership and marriage and in the agricultural value chains. However, this is likely to be challenged by men and cultural norms that challenge women's inheritance of land. It is important that funds are set aside to carry out the sensitization of these issues and includes, taking a bottom up approach (Counties, MCDAs, stakeholder and ministries). Such activities will drive change within local communities in values and attitudes with regards to asset ownership, and result in economic benefits.

5.4 Challenges of Implementing Agricultural Policies

There is a strong contradiction on policy action with regard to what is said and written on paper and what is feasible. The implementors of these policies (mostly government and local administrations) do not have the resources to deliver the policies to the levels needed (Mdee et al., 2020). The lack of clarity with regards to agricultural set goals in production and sustainability, hamper agreements between policy makers and implementors, as bench marks are predominantly set by donors and NGOs. The narrative of agricultural policies is based on driving modernization and profitability, but in the absence of sustainability. The use of state sponsored support to support production of crops like maize only seek to discourage the use of eco-friendly means. This is further connected to the relationship of power structures and electoral processes. Secondly, making agriculture less diverse.

The call for this is pushed by the opportunities the ruling class seek to gain from state support, through an increased number of middle class and ruling class amalgamation of land, as well as support from NGOs that support large scale land ownership. One might therefore assume that this drive of industrialization will seek to have the small-scale farmer work as a labourer, however there is no indication of any development. The failure of policy is evident at different levels: Worldwide policies on increased agricultural production, and climate mitigation are vague and set in a manner that raises questions on normative grounds. The policy discourse then adds to the State and local political landscape, which when added with the lack of implementation, producers and climate change, encourages the politically connected to amass resources and at the same time drive and increase a number of NGOs, who seek to compensate for the lack of state support with regard to skills and hence end up creating sections of donor driven interventions. Where the presence of agricultural support is present, then the ruling class seek to be part of it.

A good example, can be seen in the Zambia, Farm Input Subsidy Programme (FISP). The ruling class use distribution committees and traditional rulers to penetrate state supported programmes. They may include gender as part of their criteria's, but factor class as part of the consideration. These examples can be applied to land policy changes, where it makes its easier for them to acquire land from poor farmers. Thus, a wealthier woman faces more discrimination, than a poor man. It is important to note that policy changes are often supported by NGOs narratives with investment by the State in National programmes reducing as access to financing for small scale farmers is often not available hence, making it hard for out grower schemes. The ruling class and business actors are supported within the law. Markets are made hard to access and are unfavourable to small scale farmers making oppression and eviction of resources the narrative, showcasing opposition to inclusion and sustainability. Strong political support will be needed if changes are to be made. So, what happens to the small-scale farmer and their crop production? What structures their ability to work with state policies that will drive their income from crop production? Critical questions for farmers that are left unanswered as those who are part of the policy process do not develop policies through the farmers' view.

5.5 Interventions for Agriculture Policies

The Tony Blair Institute (2018) report on government reforms in the agricultural sector argues that the transformation of the agricultural sector by government leaders is hard and will require leaders to spend time fixing challenges within value chains. The focus of transformation rests in challenging political parties that take ownership of resources for their political standing by appeasing their supporters, thus, making it hard for them to concentrate on one issue as their attention is drawn on to multiple issues, they cannot ignore. The political class needs to put time and energy in building support for their goals, by using strong arguments with those who hold political power—'power owners'.

The power owners have the ability to control policymakers in government. The report asserts that for one to be politically astute, one needs to ensure the problems faced by power owners are sorted, and bring about visible economic changes for those on the ground. To get things done for example, one needs political owners of soya products to support the current government and the soya sector with local value addition, and support local business that make good returns locally with soya made products and sway soya product importers to support local processing (Said & Vencatachellum, 2018). Reducing the gender gap will reduce poverty and improve food production for Africa's increasing population. By supporting women in agriculture and introducing policies that close existing gender gaps in African agriculture can have positive benefits for women and their households. Changes in gender relations and men's support of women in the agricultural sector are seen as critical in helping women increase their production capacity (Agholor, 2019). African women are the missing link between the continent development agenda and the agricultural sector. For this, to be restored there needs to be support from State governments in ensuring gender equity is present at all levels, and that African women are empowered in the sector so that they can increase crop production, have better livelihoods and contribute to their continents growth.

However, for this to be done, there needs to be a clear understanding by international organizations about what they say they will do, what they think should be done and what they do. This is because they cannot make policy changes that affect countries directly. To have gender equity for instance the FAO Committee on World Food Security needs to affirm that women and girls have tenure rights and access to land and that they have legal means that give them sufficient protection and laws that recognize their rights and are implemented and enforced. Governments need to pursue legal services and other services that are critical in protecting women's land rights and remove any and all forms of discrimination on women's abilities to own, inherit and sell land (Amenyah & Puplampu, 2013). The perpetual problem is that States and NGOs presume that men are the only ones who work on the farm on a full time basis, and that services related to farmers should be directed to men as the head of the home. Secondly, that global institutions can only work within the settings of State agendas. Thus, in such examples where States development agendas are confused and agendas not clear, the institutions will not be able to drive the focus of improving women's needs in the agricultural sector. However, when State and NGOs agendas align, a lot of progress is achieved. The FAO in its 2010–2019 strategic objectives has sort to promote equal access, and employment opportunities for both women and men in rural communities. Gender mainstreaming is at the center of FAOs policies and programmes, and is not centered on 'women's issues'. Countries, where this has been successful and where this has been implemented by both FAO and other agencies are: (Lesotho, Gabon and Ghana) where agenda conditions supported implementations. There is also scarcity of political leadership and skills in several African countries when it comes to gender issues in agricultural development. International organizations, (FAO, World Bank, AfDB, IFAD) have developed interventions by supporting research projects. Through these they have developed 7 tools for analyzing gender and researching on gender and development. They have also supported training and created awareness on major gender issues. Through these ventures they have been able to stir State governments and policy makers. However, for these ventures to succeed there must be political will from political leaders and institutions (Amenyah & Puplampu, 2013).

Leaders drive national development within their countries and can be the driving force for policies that can change the sector. African countries have been governed mostly by men and the recent ascension to power of African women is proof that leadership can drive change. The rise to power by Sirleaf Johnson of Liberia and Joyce Banda of Malawi is an example, of the required leadership that can address the needs of women farmers in Africa. For effective change to take place women leaders also need to have men as allies to support them in making effective change. In addition, there is a need to build the skills of those working in institutions for the development of gender relevant policies for agricultural development (ACBF, 2012). Specific departments were set up in agricultural ministries in several African countries during the implementation of structural adjustment programmes in agricultural institutions in the 1980s (World Bank, 1989, 1981). Due to lack of leadership, most of these departments were not able to deliver on their mandates. Thus, national and continental institutions need to align with the needs of the sector so as to achieve desired goals by setting up agendas that are aligned to national and international contexts. The needs of each country is different, it is critical therefore that specific issues related to each country are addressed within the wider national setting.

The AU's stance on certain continental issues has had an impact on certain national fronts of countries on the continent. This has opened up an avenue for championing the needs of African women in agriculture. The development of such a policy will not only act as a driver for the continent, but will also drive countries to prioritize the same within national contexts, as women contribute heavily to the sector (Amenyah & Puplampu, 2013). The real change for African women, has been initiated by women in rural communities who have seen the need to fight hunger and poverty. The likes of the late Prof. Wangari Maathai are an example of this grass roots leadership and its impact through the Green Belt Movement via tree planting and large re-greening efforts throughout Kenya. These efforts have led to change in environmental eco-systems, climate adaptation and poverty reduction. The noble laureate changed the lives of many rural women in Kenya, by giving them hope that they can be the change they seek by tending the land and improving agricultural production and their lives (Glazebrook & Opoku, 2020). Women are agents of change, and change is possible when governments and policy systems support women by appreciating their knowledge instead of viewing them as farm labour. For instance, an initiative in Buloa Subdistrict of Makassar City, reported women who were part of the initiative as being educated on the needs of women making decisions and building their capacities and knowledge and the importance of inclusion in project planning. These experiences are not based on capital but on market factors that enhance labour conditions and outcomes. Women thus showcase change by focusing on capitalist goals with their caring labour practices for the family and community through collective and practical effort. Sustainable systems that support care have the ability to change global inequities between the North and South that drive hunger and poverty. Women's approach to eco-system services is well rounded as they view people as part of the system, rather than as bosses, and are able to adapt to local conditions, though they generate income and livelihood through their work, their main aim is not just economic growth. They are looking for ways and means in which they can provide for their families and other women as well as their communities with a view of ensuring sustainability. The disparity between local economic solutions and women is the distinction between capital and care in that, labour and exchange aimed for profit and labour and exchange or community support. Many of the world's farmers are women who grow food for their family's consumption. Women's agricultural practices warrant attention and understanding.African women have over the years developed diverse farming methods like crop rotation, fertilizer production from animal waste to keep land productive as they have difficulties in language culture and finance, due to their inability to relocate. Despite the challenges that they face, women are excellent managers of their environments. However, due to time constraints, and because of subsistence farming, their drive towards food production can be better if they work in groups and thus can succeed where men have failed, as they assist in the policy and practice of sustainability (Glazebrook & Opoku, 2020).

5.6 Agricultural Gender Responsive Policies

Environmental shocks and challenges, are increasing the gender gap and reducing efforts to achieve sufficient food production. It is of the utmost importance to reduce these challenges, by developing innovative technologies that can be easily adopted and meet the needs of women farmers with the view of improving their agricultural outputs. In addition, by developing and involving women in the development of gender responsive policies, we can have equitable and sustainable food systems (Odongo, 2023). State policies can support or hinder female farmers growth in the agricultural sector. Thus, having polices that address issues in the sector and support women inclusion can help increase women's involvement. This can be addressed by four priorities as discussed below.

5.6.1 Women in Leadership

There is a positive light in the environmental management landscape as the link between women's leadership and sustainability grows. This has been attributed to the strong impact that women have had in changing mindsets, by including inclusivity, equity, and sustainability. This thus, showcases the importance of having women take charge on issues affecting the environment during our time. Women use different leadership styles, that showcase their empathy and their adoption of using team spirit that helps them take part and be part of inclusive decision making. These principles call on the development of solutions that can stand the test of time and involve individuals from different backgrounds, disciplines and generations (Balabantaray, 2023). Women are affected by different issues and it is incumbent upon them to make changes on the issues affecting them. However, change is not a one step journey, but one that needs time to refine and present itself anew. The world today represents, the impact of change over time. It is therefore important that women be part of the change within our food eco-systems, by either acting as caretakers, managers or guardians, so as to ensure the environment is safe.

The reason, behind this is that if they don't, they will suffer the most. Climate change is not mitigated by taking a back seat. It is the courage to take things head on as shown by women that reflects the ever-changing nature of environmental issues. Women have shown great leadership in taking charge and responding to diverse challenges that come along their way. They have done this by managing and merging economic and environmental factors with the goal of achieving economic growth and environmental sustainability. Highlighting the importance of integration where two entities thrive when together, rather than when they are apart (Balabantaray, 2023). The complex outcome of laws, education and culture highlights the far-reaching lines of women in leadership. Women play a crucial role in changing systems that impact generations. They do this by steering policies that challenge multifaceted issues and act as mentors and teachers to the next generation (Balabantaray, 2023). The implementation of inclusive, equitable and empowering policies that support women's efforts in building sustainable ecosystems is important. These ideas are key in encouraging women to take up leadership. Thus, recruitments need to ensure guidelines that are gender sensitive and support leadership for women. This can be done by building a stream of girls in STEM education who can then be environmental leaders. By using coaching and mentoring programmes to champion girls to take up environmental fields through the development of networks where leaders can gather to share knowledge and build on their goals.

It is therefore, important that environmental policies consider barriers, opportunities and challenges that both genders are likely to encounter. Budgets need to be reserved for women working in environmental programmes and their achievements recognised by giving them the necessary skills in mitigation, advocacy and negotiation. Improving women's leadership skills is critical in building the agri-food systems in Africa. Thus, States have an important role in helping to propel women to take on the leadership mantle, by giving them opportunities to represent their citizens. It is also critical to train men to be ally's and to acknowledge the barriers and enablers that support women in leadership. By giving both men and women the right tools when they take up leadership and having them well represented in gender intentional-strategies and policies as critical decision makers (Odongo, 2023).

5.6.2 *Access and Communication*

Second, the manner in which information about women is shared needs to change, so as to ensure information is shared widely. Mobile and digital access is still one of the most common forms of communication that is affected by gender gaps as cited by the FAO. By using communication theories that reduce gender gaps, communications gaps can be reduced. The use of these theories must therefore note the abilities of those using and accessing the information, so as to ensure the information is understood, by using clear language that is easily understood by women (Odongo, 2023).

5.6.3 *Finance and Land Rights*

Third, policies that seek to improve women's abilities to increase agricultural production, need to be focused on the entire agricultural value chain: (agri-finance, agro-processing, access to input and research). Women globally own less than 15 percent of land. Thus, to address women's ability to access finance, policies must be put in place that seek to improve women's rights to power and tenure. This approach therefore needs to be holistic, so as to promote gender responsive policies as gender equality stems from systematic issues (Odongo, 2023).

5.6.4 *Strategic Partnerships*

Fourth, partnerships play a critical role in building and promoting sustainable farming practices that can improve livelihoods, and further facilitate gender responsive agricultural practices and policies. For example, EIAR Institutional Fellowship was launched in partnership with the Ethiopian Institute of Agricultural Research (EIAR) and the African Women in Agricultural Research and Development (AWARD) in 2018, with the goal of partnering young, newly recruited female African researchers with senior researchers as part of a structured mentoring programme. The programme not only focused on intergenerational learning, but also contributed to an institution wide culture of mentoring young talent. A set of interventions designed by the EIAR-AWARD partnership will assist in the growth of leadership and research skills of women researchers and through advocacy by senior leadership around gender-responsive research. Thus, encouraging the systematic integration of gender in systems and processes that are grounded in institutional mechanisms (Odongo,2023).

5.7 Filling the Gender Gap in Agriculture

One of the biggest challenges faced in policy implementation is the lack of effective awareness on what policies are and their impact. It is critical that gender responsive policies need to be measured so as to track their impact after implementation. The absence of this leads to poor funding, low commitment and implementation. One critical indicator for this was the need to have actors consider interconnectedness of existing barriers and the underlying causes of gender gaps as mentioned at the 67th Commission on the Status of Women (CSW67) side event (Odongo, 2023). With this, it is important that policy makers have a unified grasp of the factors exacerbating existing agri-food policies. By mobilizing work to world theories on gender equity and empowerment global initiatives such the World Committee on Food Security can serve as reference points. Thus, helping to reduce gender gaps in agri-food systems. The Gender Responsive Agriculture Systems Policy (GRASP) fellowship aims to spar changes in policy that will encourage gender equality and support women's empowerment in agri-food systems and is funded by the United States Agency for International Development (USAID) and AWARD. This fellowship seeks to have an array of female policy professionals who are confident and resilient in highlighting and leading policy changes that better food security and increase livelihoods. Policy practitioners with these skills are able to guide decision makers on agri-food policies that decrease women burdens in accessing productive resources. Furthermore, it creates a drive towards inclusive agri-food systems (Odongo, 2023).

5.8 Agricultural Politics

Environmental management is composed of laws, methods and behaviours that define how power is used across all stages of environmental policy. It also includes other factors, such as participation, accountability, effectiveness and coherence. The representation of women in environmental management is low. This is due to the fact that they are poorly trained if trained as policy makers and lack experience in planning and environmental law. Even worse, when they do have the skills they are often cast aside from the key decision-making institutions that have policy making capacities at national, regional and international levels. Women are further underrepresented in financial institutions where environmental decisions are made, thus adding to the problems of coordination between States, NGOs, and institutions that address environmental issues. Furthermore, despite the increasing numbers of women NGOs in this field their partnership with key decision-making organizations is limited (Mukherjee, 2013).

Food has become a central and a critical talking point in international circles, and amongst global, State and local actors. Food intersects with a lot of matters (production, reproduction, distribution, consumption) and between global, national and local actors (Leach et al., 2020). Food also raises a lot of questions with regards to different sectors (economy, state-society relations, and environment) and has a connected relationship to (personal, social, cultural and bodily status and identity). This relationship enables food to be part of an integration of networks that give vital insights into a range of current developmental issues. On the other hand, food also draws on politics, and asks questions related to food systems and how they developed and how they change or can be changed, and who profits or loses when power dynamics play out among the different sets of actors. What is clear, is that regardless of issues at hand, food is a highly charged topic of discussion around the world.

According to Mills (2018), the potential of African land has been hampered by those who own large tracts of land. For instance, 35 out of 48 Africa countries were net food importers by the end of the 2000s. Africa has halved its agricultural exports to under 4 percent since the 1970s. In addition, one-fifth of the continents economic output was attributed to agriculture. Two-thirds of rural women who live in the continent are dependent on farming for their livelihoods. Women who live in the rural areas and are dependent on farming for their livelihoods and are two-thirds of the population. It's therefore not hard to see why the agricultural productivity of the sector is low on the continent (Mills, 2018). Mills (2018) attributes this to the high taxation that is meant to fuel Africa's industrialization drive and the bad leadership that has added to the poor performance of the continent. AGRAs 2018 report on Africa's Agriculture Status investigates (AGRA, 2018: 35) 'why political will to support agricultural transformation in sub-Saharan Africa has remained limited despite democratization.' Credibility has been identified as the problem according to the report as the electorate lack sufficient means of appraising the performance of elected officials, as they frequently pick candidates who are of the same ethnic origin ('identity-based voting') or alternatively use local traditional leaders. Poulton and Chinsinga (2018) assert rural traditional leaders are used to promote a culture of support. This method of support or loyalty seeks to boost their ability to own the political narrative as the electorate are not learned.

Political commitment is lacking from elected leaders and threatens the prospects of increased productivity of the sector as it only seeks to support the narratives of the elected officials by having growers fight for the meagre supplies that they may receive. Changes in household farming can only be visible if there is political commitment according to the 2018 report. Furthermore, industrialization within the continent is minimal due to poor economic growth. The highest earner of foreign currencies in the continent sit within the key sectors of minerals and agriculture, supported by donor aid and limited leisure travel (Poulton & Chinsinga, 2018). 'To achieve a successful agricultural transformation, it is not only essential to develop sufficient political commitment to address the market failures and the governance challenges outlined, but it is also important to sustain political will over time' (AGRA, 2018: 25).

The absence of government support could explain the reasons why past African agricultural success such as improved maize varieties or the 'cassava transformation' in the West and South of the continent were not supported (AGRA). The agricultural sector was not prioritized in the 1950s and 1960s and was not regarded as critical for economic growth after independence and therefore not followed upon (Umbadda & Elgizouli, 2018). The focus at that time was on government sponsored industrialization that would be driven by the transformation of farming communities to modern industrial centres.

Poverty alleviation through agriculture was least concerning as economic growth was the main priority. The driving force towards this was that job creation would increase economic growth and hence, tackle issues of poverty and thus improve health and education. However, this changed in the 1970s, when the focus shifted to improving the agricultural sector with the view of alleviating poverty and improving community development (Umbadda & Elgizouli, 2018). These programmes were hampered by strict economic policies that were not sustainable and had no funding from donors. Thus, making the projects expensive and unsustainable as most of the projects were in the pilot phases. Masaki (2018) argues that donor funding on projects only seeks to showcase the deterioration of governments in managing their affairs.

Donor aid gives leaders both (democratic and authoritarian) additional money to use on political loyalists and supporters and consolidate power. This funding is critical in shaping political outcomes in sub–Saharan Africa with regards to receiving favours and buying loyalty. Others view aid as giving leaders more funds to share that will enable them buy loyalty and support (Knutsen & Kotsadam, 2020). However, there seems to be a correlation between aid before ballot casting and current political holders in Africa. This suggests that political leaders are using aid to fund themselves to remain in power (Briggs, 2015). Democracy in Africa is still very young, with uneven playing fields that are used to transport and support loyalty rather than critical reforms (Hickey et al., 2018). The lack of clear policies and manifestos is a strong indication of their inability to make critical policy developments that will bring about transformation. It is for this reason that politicians use senior political figures to build their support and win elections.

Policy makers in the African continent need to ensure that they are able to bring forth leaders and change makers, who align with key strategic development goals, rather than having those solely focused in ministries of agriculture. It is critical to have leverage with decision makers who are Presidents, ministers of finance, trade and industry, senior government officials and legislators, who can drive meaningful change and support agricultural transformation. These leaders need to have leadership perspectives that speak to the same level of thinking as that of their President, Ministers and senior government officials. In addition, they need to have an understanding of the global perspectives, and the impact of these perspectives with regard to their own countries, so that they are able to navigate through the political landscape, patronage networks, execution challenges, security concerns, and economic and political pressures. Transformation, can thus take place when such approaches are used in both local and international advocates as well as Presidents, ministers and other key decision makers with the view of aligning to key economic policies that can transform the sector and drive economic prosperity (Said & Vencatachellum, 2018).

5.8.1 Women in Politics

Women empowerment is a powerful tool for development as argued by Nobel laureate in economics, Esther Duflo (Duflo, 2012). Prior to that, Nobel laureate Amartya Sen, argued 'that development is driven by the institutional environment that allows 'freedoms' to flourish, where people have 'agency' to follow their desire for doing well, and termed 'capability' as an approach to development (Sen 1999).' He reveals economic growth as being constrained by the lack of agency among women, who are mostly poor. The gender and development narrative has been argued and challenged on different fronts. First, on how women's advancement can commence with the need to change social justice and the reasons behind why women remain subordinated to men. Others seek to showcase women as carrying heavy loads of limited progress (poverty, disease and violence). The truth of the matter is that female empowerment is not just an inner motivation, but a valuable asset that encourages development, by supporting the redistribution of resources to towards skills that will encourage political and economic growth (Hornset & De Soysa, 2021). Women's empowerment is a valuable asset that improves their mental health, agency and self-worth.

Gender equity, is one of the SDGs that needs to be achieved not as an instrument of value, but because it is right. However, this is not the reality on the ground, as 27 percent of management positions around the world are filled by females, up to 104 States possess legislation that work to prevent women from participating in social, political, and other spheres of public life thus affecting their lives. For instance, women represent two-thirds of illiterate people (World Bank, 2018). Engaging, educating and empowering this group of women is critical in empowering them socially, and goes to improve their chances of having a better life in the wider spectrum of human capital development. Women are not a group to be identified within one homogenous group, but rather are individuals who have unique skills, attributes and identities. Thus, women empowerment and empowerment of society are totally different. Gender therefore, stands as a critical part that can be used to address exclusion within allocations and different social segments with the goal of ensuring equitable distribution of public goods by States to show case the distribution of power in society. To free women from social norms that hinder their agency is a great and powerful way via private means that are linked to household decision making and through impacts on public institutions.

5.8.2 The Green Belt Movement

A few traditional African societies, do not involve women in the decision-making process and the women in these societies are not complaining. However, others view opportunities granted to them by men as their right without any query. In today's society, change has dramatically resulted in African women having received a Western education. Having undergone this type of training for years, women have learned that they cannot be set aside in any facet of life, like in certain communities in Africa. To these women, taking part in decision making is their right and they can't be denied by men (Egbai & Uduagwu, 2018). This is a good example of how the African feminist (women) view the fundamental form of control by men. Men's control of ecological and political dimensions is visible in certain African countries, thus, showcasing patriarchy. It is within these settings that a unique form of feminism, known as ecofeminism arose. It seeks to raise the voices of women in nature and politics. The cause has been championed by Wangari Maathai, Sophie Oluwole, and many others. Ecofeminism is a model for political messaging in the African space. Its main objectives are to fight deforestation, desertification, water crises and environmental degradation and encourage women to be change makers via politics. Thus, the Recognition of women's contribution (Voice of Women Contributions') stands out in modern African communities, by challenging ecological harm, female leadership, patriarchy along with other forms of domination within the society. The late laureate and Professor, Wangari Maathai, was a well known ecofeminist who was born in Nyeri town in Kenya. Her major contributions to the ecofeminism cause have encouraged women to take up leadership positions in politics from the grass roots to national levels by establishing theGreen Belt Movement (GBM), empowering women to build connections as well as self-confidence as they gear up for political action in Africa, by developing groups of women who work together to protect and conserve the environment, and are actively involved in environmental politics, while highlighting the importance of being learned in Africa, and the important role culture has in conserving and preserving the environment within political ideologies (Egbai & Uduagwu, 2018).

How was this achieved? The GBM was a key initiative of building women's connections and their self-worth. Thus, the establishment of the GBM was a way to build an eco-system for African women to connect and build their self-worth, regard and respect. The key objective of GBM was to increase women's self-confidence through civic education. Using this movement, she was able to empower women in communities and mobilize them for activities (2004). She writes about this journey in her publication titled 'Replenish the Earth', where the impact of her work was focused on the GBM goal of saving trees and the environment, and how this concept was shared among African women. In her quest, she realized the need to have women find their true selves and the need to expand open spaces so that women could be heard (2010). The establishment of GBM, was an avenue out for some African nations. Thus, she was able to grow strong relationships among women and some political associations (Egbai & Uduagwu, 2018). She did this by using, conferences, newspapers, workshops and one on one communications with people and within groups, to the point where women's connections moved beyond tribes, and women were able to work together for the common objective of protecting and conserving the environment as well highlight political issues in the continent. Thus, women's team spirit arose. She used this awakening to encourage women to participate in politics while conserving the environment. She explained this as working together in teams and then indicated that working in groups is not enough and that one needs to practice what they have learnt from the group to grow within their community. Thus, one must internalize what he or she has received from the group and have the courage to act alone or with others. However, for this to work effectively, individuals needed to have built their self-worth and self confidence (2013). She noted the importance of having African women work together for the betterment of the world and that participating in politics was just one of the ways, not only the way but the best way of becoming politically active. Thus, the grouping of people together towards a particular cause is a good case of 'political action' in the wider perspective. In addition, the GBM is an excellent avenue for women to share their views on environmental management (Egbai & Uduagwu, 2018). Kusher argues that the organization serves a programme that recognizes women's connections with environmental degradation and poor governance, and that by encouraging women's participation in politics, GBM addresses the core issues affecting the environment and society by teaching people about the intersections of poverty, racism and capitalism. Her third and most significant contribution was her ability to gather women through the GBM and encourage them to take up political leadership. This challenge, has given rise to more women participating in political activities linked to electoral processes, parties and asserting their influence in government through policy making (Egbai & Uduagwu, 2018). The involvement of women in active politics came about in the 1990s when women associations like GBM, community improvement associations and NGOs supported and prepared women for political work. In this roles, women voiced strong opposition to graft and ethnicity that were present in the governance structures (Tripp, 2004, 144, 151–152). The fourth contribution, was her ability to showcase the importance of educating African women. Presbey (2013) asserts that educating girls will raise countries national incomes exponentially, as a result of this education one can become active in politicking. Her fifth contribution was her role in showcasing the importance of culture and its critical impact in environmental conservation. She argued Africans needed to preserve their traditional culture, because the moment they stopped doing so and adopt Western forms of culture, they would be stuck with 'gods of commercialism, materialism, and individualism', which drains the body and soul (2009) (Egbai & Uduagwu, 2018: 112).

Cultural practices in Africa are used to conserve the environment, for instance nature-based worship of animals, trees and mountains. African women are critical players in environmental conservation, as they often communicated traditions, such as telling children where not to farm or where not to cut wood. In the Nigeria State of Benue, the praise of the green snake is upheld by women. African traditions have always had a role in supporting women to take part in leadership, as seen among the Igbo-in Nigeria where women assemble in groups like 'Umuada' (daughters of the land) and 'Ugosimba' (those married to the community). It is through women's participation in these groups, that they are able to take up leadership and engage in the political space (Egbai & Uduagwu, 2018).

5.9 Conclusion

Women need to be at the forefront in developing policies that affect them in the agricultural sector. Their absence will lead to the development of policies that will seek to bar them from progressing. Furthermore, policies need to take into account the different social and cultural norms that affect them. This is due to the fact that women are affected by a barrage of issues depending on where they reside. Thus, National, State and local contexts need to be taken into consideration. For instance, local actors were not able to connect transformative gender solutions about inequality that would challenge local power dynamics, largely because the policy actors had framed patriarchal social-cultural norms as barriers to gender equality in agricultural and climate change development and where norms were largely grounded in proposed 'local' solutions (Acosta et al., 2021). Local policy actors have often shifted focus away from the root causes of gender inequality by politicizing gender issues. This politicization deflects attention from the deeper structural inequalities at play, reframing the problem as merely a lack of knowledge about gender concepts or policies. As a result, even when gender policies exist on paper, they may not be implemented effectively or integrated into broader policy frameworks. Ensuring women's active participation in policy development is crucial, as it provides critical insights into addressing challenges such as access to land, financial inclusion, and training. Finally, policies that actively support women in agriculture will not only encourage greater female participation in the sector but also position women as key stakeholders in shaping agricultural policy.

CHAPTER 6

Women's Power in Irrigation: Resilience and the Ecofeminist Link

6.1 Introduction

Agricultural technology and innovations have made great strides over the past 50 years (1960–2010). However, feeding 9-billion people in 2050 and beyond will require the intensification of agricultural production by as much as 40 percent while also maintaining ecosystems and planetary integrity. Limited arable land and water for irrigation poses great challenges for humanity in reaching this goal. We therefore need to move beyond the business-as-usual scenario (International Water Management Institute, 2014). This chapter explores the role of women in irrigated agriculture in the Ahero Irrigation Scheme in Kisumu County in the Western region of Kenya. Irrigated agriculture is seen as one of the major means through which food security may be improved in Kenya. Women face various challenges when it comes to taking up agricultural activities: from insecure access to land, credit, agricultural inputs and lack of proper agricultural training. This chapter seeks to highlight the role of women in irrigation farming in Kenya, discuss the history of irrigation in Kenya, the history of the Ahero Irrigation Scheme and the challenges women in the Ahero Irrigation Scheme face concerning irrigated agriculture within the themes: women and land, women and water, financial resources and assets, technology and education and agricultural politics.

M. O. Jimoh and S. Narsiah, *Unveiling the Power of Women in Irrigation*, https://doi.org/10.1007/978-3-031-94174-0_6

6.2 Women and Irrigated Agriculture

Irrigation is becoming an important trigger for agricultural growth in sub-Saharan Africa with both women and men adopting small-scale private irrigation in ever large numbers. This trend in farm-led agricultural growth promises to enhance women's productivity and well-being (Nicol et al., 2015). This is because irrigation is a major factor in agricultural intensification. Furthermore, irrigation projects have helped to increase food security around the world, particularly in developing countries (Mateo-Sagasta et al., 2018). In many societies, men and women have different distinct roles in agriculture and food production. These roles have been influenced by social and economic contexts. Women are increasingly taking on greater and more varied responsibilities in farming, especially in the field of irrigation, managing labour and procuring inputs (CGIAR Research Programme on Water, 2018).

In Kenya, for example, women make up between 42 percent and 65 percent of the agricultural labour force according (Diiro et al., 2018). They argue that despite women's important role in the agricultural sector, there is a lack of empirical evidence that shows that they lag behind men. Gender inequalities continue to persist in respect of and access to, productive resources such as land, livestock, labour, education, extension and financial services and technology. In addition, cultural norms and traditions restrict women's ability to inherit land and contribute to the widening gaps of farmland owned between men and women.

Gender equity in irrigation is important if agricultural production levels are to remain high in irrigated farms. This is because empirical evidence of gender differences in farm productivity shows that women do not have the same access to resources as men. In addition, if women are accorded the same opportunities as men they are more than likely to increase their yields by 20–30 percent. Productivity gaps exist in favour of male-headed households, with gaps in the region of 4–40 percent. Household economic surveys usually interview the 'head of household' which often leads to a male biassed perspective and thus misses the complexity of intrahousehold relations (CGIAR Research Programme on Water, 2018). The CGIAR Research Programme on Water (2018) asserts that many studies have shown that well-intended reforms aimed at providing women with more opportunities often fall short because the contextual social dynamics have not been well understood. This is because well-meaning development programmes often fail to engage with the structural barriers that women face in accessing irrigation and other farm resources. It is therefore important that we understand the challenges women face in the irrigated sector and how best they can be addressed. However, before discussing this, it is necessary to present a brief history of the irrigation sector and the role it has played in agriculture in Kenya.

6.3 History of Irrigation in Kenya

According to Kisang and Yator (2017) irrigation has been in Kenya for over 400 years and has existed for many years along the River Tana, Keiyo, Marakwet, West Pokot and Baringo regions. Furthermore, rice irrigation activities have existed along the river valleys around Kipini, Malindi, Shimoni and Vanga, since the early nineteenth Century when slaves were used to construct rice schemes. In addition, Asian workers building the Kenya-Uganda railway also started some irrigation activities around Makindu and Kibwezi.

'Irrigation schemes in Kenya can be categorized into small-scale schemes, large-scale private/commercial irrigation schemes and public schemes. Large farms occupy 40 percent, while small-scale irrigation schemes account for 42 per cent, and the remaining 18 per cent belong to government-managed schemes (Muema et al., 2018: 2).' They noted that the main irrigated crops grown within these schemes were rice, vegetables, coffee, fruits, sugar cane, cotton, maize and horticulture. Kisang and Yator (2017) indicate that the Kenyan government established the National Irrigation Board in 1966 to manage and run its irrigation schemes. In addition, according to Kisang and Yator (2017), the irrigation board was also mandated to open new schemes and thus create employment opportunities for detainees, and the resettlement of landless people. Furthermore, they engaged with the farmers through tenancy agreements. This was however seen as a downside as the management of the schemes decided what crops were to be grown.

'78 percent of the total area under rice cultivation in Kenya is in irrigable ecosystems which are managed by the National Irrigation Board (NIB)' (Atera et al., 2018: 66). The schemes under this care include Mwea, Bura, Hola, Perkera, West Kano, Bunyala and Ahero. Rice production in Kenya dates back to 1907, when Europeans introduced it along the coast. Rice is perceived as a rural cash crop for the rural population, and has been seen to be appreciating over the years (Atera et al.,

2018). In view of this let us have a brief history of the study site, to better understand its history.

6.4 History of the Ahero Irrigation Scheme

According to the NIB (2018) website the Ahero Irrigation Scheme is located in the Kano plains between the Nandi Escarpment and Nyabondo Plateau in Kisumu County, Muhoroni sub-County in Kenya. The scheme started as a pilot project to explore the feasibility of irrigation in the Kano Plains. Construction of the scheme started in 1966 and the operations of the scheme began in 1969. It has a gazetted area of 4,176 acres, with 2,586.5 acres under irrigation. The scheme has 570 farm holders, 2,000 farmers and 20,000 dependents. The source of its irrigation water is River Nyando. The scheme uses basin irrigation, with rice the main crop in the scheme. The varieties of rice are 90 percent Sindano (IR-2793 & ITA 310), 5 percent aromatic (Basmat-370) and 5 percent hybrid (ArizeTej Gold &Arize 6444 Gold). Other crops like soya beans watermelon, maize, tomatoes, sorghum and cowpeas are also grown within the scheme.

According to the Kisumu County Urban Sustainability Review (2017), agriculture is the economic backbone of Ahero. It employs about 70 percent of the population. According to National Irrigation Authority report (2019) by the Ministry of Water, Sanitation and Irrigation, most farmers in the scheme did not adopt modern cultivation farming. Women farmers are particularly affected as they spend a good amount of time attending to household tasks and have limited time to take part in technology training. Furthermore, women are excluded from land inheritance. Land in the scheme is registered under the name of male farmers. Men in the scheme do not share the skills gained with their partners. The lack of secure land tenure rights makes it hard for women in the scheme to collect loans. Thus, these scenarios exacerbate poverty among the women in the scheme, as they are not part of the decision making process even though they performed key tasks in agricultural production. Male farmers who did not participate in cultivation made major decisions concerning cultivation. Female farmers were mainly engaged in postharvest activities of threshing and winnowing.

The irrigation subsector is currently grappling with numerous challenges (Mugambi et al., 2017). This is echoed by Muema et al. (2018: 2), who note that 'irrigated agriculture is faced with challenges such as inefficient water use, low returns on investment, poor management and suboptimal operations.' Mugambi et al. (2017) noted that these challenges included low rates of irrigation infrastructural developments, inadequate funding for development and investment in both the public and the private sectors, as well as poorly developed marketing channels within the irrigated commodity value chain. In addition, they asserted that policy, legal and regulatory frameworks have not been reviewed to reflect the emerging operational and socio-economic realities. Irrigation scheme development, according to Van Rooyen et al. (2017), requires huge engineering investment in constructing storage, diversion infrastructure and the associated canals. In addition, irrigation demands the human capacity and institutional arrangements to manage the water, ensure equitable distribution, collect water payments, carry out repairs to maintain the scheme and manage disputes. It is for these reasons that irrigation cannot be a subsistence activity. Furthermore, the linkages to input and output markets and the ability to secure sound technical information, credit and other services are critical. Moreover, irrigation development displaces some farmers and imports new farmers, leading to social and political tensions over land tenure.

6.4.1 Inhabitants of Ahero

The communities in Nyanza and western Kenya comprise of two linguistic families, the Bantu and the Nilotes (Ndeda, 2019). The Bantu of western Kenya comprise of the Abaluyia, Abagusii and Abakuria, while the Nilotic group consists of three groups: the river lake Nilotes who comprise of the Luo who practice fishing and reside near the shores of Lake Victoria; the plain Nilotes who comprise of the Masaai and the highland Nilotes who comprise of the Nandi, Samburu, Turkana, Teso, Njemps and Nubi who practice pastoralism. The various clans in the Luo locations of Kenya have diverse historical origins but have a dominant culture and a common language (Dholuo). The Luo came from Sudan several centuries ago. According to Ogechi (2019: 115), '87 percent of the inhabitants in Nyanza are Luo speaking indigenes.' 'Although fishing has been the main economic activity in the region, irrigated agriculture is taking off in the region due to the decline of fishing in the last two centuries due to overfishing and the spread of predator fish species such as the Nile Perch (Hebnick, 2019: 13)' (Fig. 6.1).

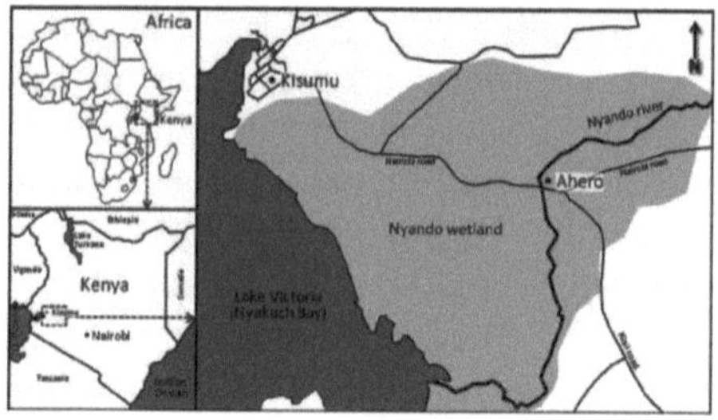

Fig. 6.1 Nyando wetland, Western Kenya (Adapted from Asking & Nilsson, 2019: 1)

6.5 The Role of Women in Irrigated Agriculture: A Case Study of the Ahero Irrigation Scheme

According to You et al. (2014: 34), ‘Kenya is home to more than 33 million people. It is rich in natural resources and yet poor and food insecure. In addition, 80 percent of the impoverished live in rural areas where 30 percent of households are food insecure.’ You (2014) goes on to state that Kenya needs to desperately find an effective solution to boost its vital agricultural sector so as to keep pace with its growing population. The Overseas Development Institute’s Report on Gender, Agriculture and Water Insecurity informs us that ‘the agricultural sector is essential in reducing poverty and achieving sustainable development goals’ (2016: 8). According to Johansson (2016: 14) ‘gender’ is explicitly mentioned in water policies in the contexts concerning countries’ domestic water developments. However, when women are viewed as irrigators, fishers or farmers, they are less recognized. Zwarteveen (1997) showed how women were not incorporated in irrigation policies. Furthermore, irrigation management is closely connected to land rights, which many women in developing countries have difficulty in accessing based on traditions.

Researchers in the 1980s and 1990s documented women’s contribution to water in domestic and agricultural contexts and that, while this was ongoing, little attention was paid to gender issues in irrigation schemes (Lefore, 2017a). The limited research on gender reinforced men’s control of land and reduced women’s input. This is because the corresponding benefits of irrigation with regard to income and produce were not received by them. This, therefore, highlighted the underrepresentation of women in irrigated land ownership and scheme management and resulted in the women withdrawing their labour from the irrigation schemes, resulting in lower scheme productivity levels. Ensuring gender equity in irrigation is important because the level of agricultural productivity of women can be the same as that of men when they have access to the same quality and quantity of agricultural inputs, potentially increasing women's yields by 20–30 percent above current levels (Lefore, 2017a). Agricultural Water Management (AWM) is one of those inputs that can increase agricultural productivity for both women and men, thereby contributing to food security and improving livelihoods. It is therefore important that most governments prioritize the development of new irrigation schemes over the rehabilitation of old ones. At the same time, both governments and development partners should acknowledge the centrality of gender in agricultural development goals. This is because achieving gender equity in the irrigation sector has become an important development goal.

However, Meinzen-Dick et al. (2014: 98) argue 'that even with recognition of the importance of women's control over natural capital, with regard to land and water, there is no straightforward path to strengthening women's rights to these assets. Statutory laws are only one (though albeit important) source of rights that many women fail to use, owing to a lack of legal knowledge and poor legal implementation. On the matter of the level of education among farmers, Hart and Aliber noted that 'the low levels of education attained by female farmers are a cause for great concern. This is because education is crucial to enable end users adopt and make optimal use of new technologies' (2015: 4). They argued that education levels of farmers will need to be boosted, especially as farmers engage in more sophisticated input and output markets. According to Fabiyi and Akande (2015), government leaders must address the issues affecting agricultural productivity. This is because agriculture is seen as playing a critical role in fighting poverty and food insecurity in Africa. Fabiyi and Akande (2015: 237) argue 'that in most developing countries, agriculture is one of the major pillars of the economy. Encouraging agriculture-based innovative ventures will help the advancement of women in rural areas. It is therefore necessary to empower rural women through agricultural policies that assist them take care of their needs and alleviating poverty.'

6.5.1 *Ahero Irrigation Scheme Data*

This study was grounded in qualitative research. Data collection for the study took place in January 2018 within the 12 blocks of the Ahero

Irrigation Scheme. Participants for the study were drawn from these blocks due to the diversity within the 12 blocks. Participants from these areas represented different (clans and kins), all are from the Luo tribe who are residing within the scheme. The study collected data using semi-structured interviews and structured questionnaires. The study used purposive sampling to select a total of 66 participants from within the 12 blocks for the questionnaires.

This is because not all residents within the blocks were present during the days of data collection. Of the 66 participants, 42 were women and 19 were men while five 'key informants' took part in the interview process. Thus, the study was able to have three women and two men, who were part of the local farming community within the scheme. One of the men interviewed is a manager within the Ahero Irrigation Scheme.

Marital Status

The study gained information on female as well as male farmers who resided within the Ahero Irrigation Scheme in Western Kenya. The scheme had 565 farmers who were located within 12 blocks within the scheme. The study was able to access 12 percent of this total farmer population. Although this proportion was low, the respondents were sampled from the 12 different blocks within the scheme. The inclusion of respondents from the 12 blocks ensured adequate spatial coverage and richness of the data that were collected as different clans were housed within the different blocks. This diversity among the respondents helped facilitate different views on farming from both genders, and from the diverse clans. The respondents comprised 68 percent females and 32 percent male. Sixty-four per cent of the respondents of the sample in the study were married, while 35 percent comprised of widows and 1 percent were widowers. Marital status within any society is important. In the Ahero Irrigation Scheme, the respondent's marital status is important. This is because without holding a strong position within the family, one is unable to access the rewards that come with marriage, in this case (land ownership). For example, one respondent noted that women have challenges in getting access to clan land, by stating that: They can easily get land after their husband...That one she can't have?... Atapatawapi- (Where will she get it?) ...when the husband is not there okay...who can give her?

Ng'uono et al. (2014: 11) indicates 'land ownership or access to other productive resources and the organization of agricultural production is influenced by cultural practices and traditions. For example, rules of land inheritance (by lineage, gender and/or culturally determined characteristics give one core access to land)'. In some communities, land is owned by a tribe or kinship group. Thirty-five per cent of the sampled population who were widows indicated that the acquisition of their land came from them inheriting the land from their husbands, as some were named as successors; and if the land was left to them as per their late husbands' wills. However, this was not always the case and some of the women faced challenges from their in-laws, as the manager of the scheme explained:

> 'First generation the father has died and the son took over, I have left my wife and sons and maybe I named my wife as the successor, you see I have moved my wife as second generation, what happens is there is a dispute as this land belongs to our father, and our father named my brothers as the dependents. I have taken over and neglected my brothers and named my children and my wife as my successor, so there are such disputes. That this land belongs to our father, it is not you, the woman who came to take over, even if you are named as the successor, you are the wife. We have these our sons, the in-laws so there is that dispute. But since you are the woman, the wife even if you look at the original file our names are there, so we are the ones who are supposed to take over the management of the farm. So it is like, can we inherit you, so that you are under us, so.... in as much as you are named the successor and you are the wife, the sons-in-law, who are the enemies will try and force that change, in as much as you are named the successor, you should be dependent like you are a child...'

Agricultural development in Kenya for women is hampered by the fact that land ownership is the preserve of men and not women (Ng'uono et al., 2014). Furthermore, women do not have the power to make decisions on land. In many patriarchal African countries, cultural customs dictate that if a woman becomes a widow, she has to remarry one of her husband's brothers.

This custom allows the woman to continue having access to land and food security. If not, she would have to leave the lineage of her dead husband. Makhado and Pelizzo (2016: 33) argued that 'this leaves women struggling for their own subsistence.' According to Makhado and Pelizzo, it is important to note that women's access to land is contingent upon their relationship to men and their marital status. Since 1 percent of the proportion of respondents sampled were widowers, it was interesting to note that the above scenarios do not seem to play out with the same magnitude. However, one respondent noted that '…ladies are not entitled to inherit land…I will divide the land among my sons'.

'The patriarchal world outlook results in men's governance over both women and nature' (Lv and Wang, 2018: 1660). They added that it was only by uniting the two together can they jointly break loose the shackles of the patriarchal world. Thus, from an ecofeminist perspective, as Samandari (2017: 5) posited, 'many societies have historically approached physical vulnerability by placing unequal burdens on women and men, leading to the domination and privileging of men over both women and the environment, for example, by traditionally granting men and not women ownership rights to land.'

Household Leadership

Forty-eight per cent of the respondents within the study came from male-headed households, while 38 percent were led by female-headed households. Leadership within every household is important and becomes critical when it comes to financial expenditure within the households. Chigbu (2019) has asserted that the role of women in household management may be a factor of women's disempowerment in some communities within sub-Saharan Africa because it places them in a dependency relationship with men who are breadwinners. However, this might not be the case in some other sub-Saharan communities where women are mostly the breadwinners. In other communities, women are seen as empowered, rather than disempowered. It is therefore important to consider the 'differences,' rather than only the 'commonalities' when tackling development challenges that relate to cultures, gender and people.

Number of Individuals per Household

The number of people within the various households within the Ahero Irrigation Scheme varied. Sixty-two per cent of households in this study stated that they had between 1 and 5 people living in the household, followed by 29 percent of households stating that the average number of people within their home was between 6 and 10, only 1 percent of the households had between 11 and 15 people, while 8 percent of the respondents did not answer this question. There are several reasons why the sizes of households were small. For example, older children may have moved out of the family household; family members migrated out of the scheme, but the study has no concrete answers as to why the number of household members ranging between 1 and 5 are few.

Age

Thirty-six per cent of the respondents in the Ahero Irrigation Scheme research were aged 60 years and above, followed by 24 percent who were between the age range of 51 to 60 years, 21 percent fell in the age range of 41–50, 10 percent between the age range of 31 to 40, nine percent did not respond to this question, while there were no respondents who fell in the age groups of 21 to 30, and 15 to 20 years. Maiga et al. (2017: 4–5) claimed that 'to make a livelihood out of agriculture, one needs to acquire or rent land and that ill-defined land tenure policies in many African countries, hinder the ability of individuals to buy or rent land.' Maiga et al. (2017) argue that one of the main reasons why the youth are leaving agriculture is because of the lack of access to land. The study asserts that the respondents who hold titles to the land are from the first, second or third generations, and have inherited the land from their grandfathers and fathers as several of the respondents noted for example, one women asserted '... I have, I got it from the father of my husband, who gave it to my husband...'.

Education Levels

Education has a great impact on farming. The study noted that only 3 percent of the respondents had attained university education. A further 9 percent of the respondents had received education up to tertiary level; 24 percent attained secondary education, 56 percent of the respondents attained primary education, while 8 percent of the respondents did not respond to this question. As Sharma and Maheshwari (2016: 61) argued, 'when training opportunities for rural agricultural producers arise, they mostly go to men, who already have some formal education.' One male participant when asked if education is a factor hindering women from becoming better farmers noted that: some women believe that farming is only for men and another stated that women lacked knowledge in farming techniques. The 'low levels of education attained by female farmers are a cause of great concern' (Hart & Aliber, 2015: 4). Education is crucial to enable end users to adopt and make optimal use of new technologies. According to Hart and Aliber (2015), the education levels of farmers in Africa will need to be boosted, especially as farmers engage in more sophisticated input and output markets. As the manager of the scheme noted:

> ...Looking at the Kenyan education system ... fortunately, our farmers at least have some basic education. It is not easy to train them...you know to be a great leader like a secretary you have to keep records of cash transactions, and they do this as a group, and it is done by the group leadership, as they have to negotiate on the price of the crop. So, ... if you lack basic knowledge then it affects....

The highest level of education attained by women in Ahero Irrigation Scheme is in primary level at 67 percent, 24 percent attending secondary school, and 2 percent attending tertiary education (attending college not University). Thirty-eight percent of the male respondents had primary education; 19 percent had secondary education, 24 percent had tertiary education (attending college not University) and 9 percent had university education. Ten percent of the respondents did not respond to this question.

Environmental education initiatives and sustainable development strategies that do not promote the full participation and empowerment of women and girls are bound to fail (Mukoni et al., 2018). Bozchelouie (2018) claims that the reason for this is that agricultural education is strongly patriarchal and rarely leads to improvements in the education of rural women. Mukoni et al. (2018) support this argument and assert that women as a group have continued to be marginalized due to the influence of patriarchal systems which often privilege men's voices over those of women. Consequently, women have not participated fully in environmental education programmes. Having programmes that ensure women's participation in irrigated agriculture is a step forward in ensuring women gain better access to agricultural skills and training.

Employment Status

Sixty-four per cent of the participants viewed themselves as farmers. This was important to note as some farmers within the scheme undertake farming as a secondary occupation. Also, some respondents stated that they were unemployed, while others were employed. Moreover, 33 percent of the female respondents stated that they were unemployed with only 58 percent of the females indicating they were farmers. Sixty per cent of the male respondents viewed themselves as farmers, 15 percent as retired, 5 percent as unemployed and another 5 percent as employed. These figures of 33 percent differ strongly when compared to those for males where 15 percent were unemployed while 60 percent stated their profession as 'farmer.' It becomes clear from these data that the higher proportion of unemployed individuals within the scheme were female.

One can argue that this group of females within the scheme 'need to be engaged' (Sharma and Maheshwari, 2016: 61). In addition, most rural farmers were illiterate and poor and were not be in a position to gain new skills and knowledge that could help them become self-employed.

Duration of Stay at the Ahero Irrigation Scheme
Forty-one per cent of the residents who were sampled lived in the area of the Ahero Irrigation Scheme for 20 years and more. This meant that a large proportion of the respondents had called no other place home. This figure (41 percent) closely resonated with those between the ages of 60 years and above, who comprise 36 percent of the respondent population. Only 3 percent of the population sampled lived in the scheme for a period of 1 to 3 years. This meant that they may be new inhabitants who may have bought land from a previous owner and have decided to settle in the scheme, hence showcasing the young age group of 31–40 years. Those who are the 'inbetweeners' who have lived in the scheme for 16–20 years are at 20 percent, while those who have lived in the scheme for 11–15 years total 14 percent of the population. The remaining have lived in the scheme for 6–10 years.

6.5.2 Women and Their Role in Agriculture

In Ahero, we note that 68 percent of those who participated in the study were women and that they considered farming as their primary occupation. One respondent noted that '...they have a major role to play, because there is some work especially in rice plantation, believed to be for women.' and another respondent noted that '...I feel women have a major role to play in agriculture because most women are always serious in water.' ... they are always determined, therefore women have a major role of ensuring the agricultural sector does not collapse as stated by one male respondent.

Men within the Ahero Irrigation Scheme recognize women's contribution towards agriculture within the scheme. This was important as it showed that some of them saw women within the scheme as important players in the agricultural sector. Johansson (2016: 14) argued that 'the inclusion of women, in many water projects appears only on the surface. Women are either briefly mentioned in water management descriptions or only given power on paper; and that even in some cases women are represented by their husbands or brothers in water committees.' Johansson (2016) asserted that women were less recognized when they were viewed as irrigators, fishers or farmers. Doss et al. (2018: 69) argued that 'the global community is mobilizing support for Sustainable Development Goals (SDG) and that 5 of these goals focus on gender equality and women's rights, and at least 11 of the 17 SDGs require indicators related to gender dynamics.' The second goal of ending world hunger explicitly mentioned addressing the constraints for women small-scale food producers and the nutritional needs of women and adolescent girls.

Doss et al.(2018) noted that this mobilization has contributed to a growing demand for nuanced and accurate data on women's contributions to food security and that despite this emerging global movement for reliable indicators, well-intentioned but statistically unfounded stylized facts on women, agriculture and the environment continue to circulate in written documentation.

Challenges Faced by Women in Agriculture

Women face numerous challenges that affect them within the scheme. One of the male respondents noted. 'They face challenges like discrimination in the scheme. Many of them are not allowed to own a farm. And I think gender equality and fairness should be among us when it comes to land ownership.' The scheme manager asserted:

> 'Mostly it is about land ownership. None of this land is owned by the farmer holder. It is trust land. But the farmer is a holder, when the land was being distributed to farmers, the licensee. In the books of the scheme many of the owners are mainly male farmers ... okay and we have what we call succession on these farms. The policy is that the person needs to name a dependent as a successor from the family. A majority of the farm holders name their sons as the major successors. That means if the man goes, the person who gets land are the sons, who are male. The wives and daughters come as dependents.'

Other respondents have noted the following: the 'lack of water for irrigation, poor leadership by the management, the government should provide more irrigation equipment; women should also be allowed in the leadership' by a female respondent; 'Women lack skills and strength to be farmers' as stated by a male respondent; 'financial problems—capital, lack of capital' a female respondent. One other respondent noted the following: 'Capital to start agriculture—we have skills and knowledge but how can we access capital? It is difficult, so we overcome this by borrowing loans from other firms.' A final response regarding the challenges that women farmers experience is enunciated in the following response: 'Discrimination—most people feel that women are weak and therefore cannot actively take part in agriculture. We normally try to get involved in any activity even those that people feel are meant for men, to show them that we can do the same.'

Farmers also stated that accessing finance was one of their biggest challenges while other challenges included accessing farm inputs such as fertilizers, seeds and other farm inputs. The challenges women faced within the Ahero Irrigation Scheme were viewed by both men and women within the scheme. Some of these challenges were similar to what had been mentioned by key authors on the subject, namely, land ownership, lack of capital and water access.

Cultural Barriers

The study sought to understand if any traditions hindered women from farming within the scheme. A few respondents noted that a few traditions impacted their farming while some felt no cultural barriers affected them. One respondent noted that: '...a lot of cultural barriers hindered women from farming.' These beliefs included; no farming on burial days or even when one is pregnant. Another, noted, 'relatives of a dead person should not go to the farm until the person is buried.' and 'a woman who gives birth to twins should not go to the field until cleansing is done. The reason for this is that it is believed that women are dirty and may cause hailstones to rain and destroy the crop.'

Some respondents contended that cultural beliefs did not have a strong impact on their farming except when linked to two major socio-cultural aspects within their communities, namely, death and childbirth. The death of any farmer within the scheme was treated with the highest respect and regard as no farmer was allowed to farm during the funeral of another farmer. It was presumed that the land shall be cursed and that hailstones will destroy the crops if the land was cultivated. The second was the birth of twins. A woman was not allowed to go to the field after birth rather the mother needed to nurse her baby during this time.

Having explored some of these challenges it becomes clear as to why some women, are not able to till their agricultural land. Farming is an age-old activity that still bears certain socio-cultural aspects that affect it as an economic activity. These aspects are not likely to be removed because they have been ingrained within communities, from one generation to the next. Consequently, it would take a lot of convincing to change the belief systems that have been passed on from one generation to the next. Also, men and women play central roles in agriculture. Some of the respondents noted this and were keen to point out that men carried out some of the key activities within farming in the scheme, including supporting the women farmers in the rice fields.

Women's Perspective on Men's Support of Their Role in Agriculture
Women within the scheme felt that they received some support from the men as noted from the statements; 'Yes, men do offer services that require a lot of energy that we women cannot do, this includes ploughing and harvesting of rice for stacking during harvesting.' And 'Yes, they help us perform some hard farm activities like harvesting the rice.' On the other hand, some of the respondents did not feel the men within the scheme were supportive of the women: '...men feel they can do all the farm activities, so they don't support women's role in agriculture.' Furthermore, some of the male respondents, felt that without them being present women would be unable to conduct agricultural practices on the farm, as stated by some of the male respondents noted that *'Without men, women cannot do farming alone..., some duties are harder and require men to do; that is, during the dry spell water issues are above women's capabilities.' 'Yes, some farm activities are not done by women (e.g., harvesting of rice, puddling) since they require a lot of energy.'*

According to the scheme manager '... in farms where the farm succession has been given to a lady family member. The output of agricultural proceeds is better because activities are controlled by a lady.' In rural areas of the developing world, women play a key role in running households and making major contributions to agricultural production. However, the inequalities that exist in agriculture between women and men make it difficult for women to fulfil their potential. Women rarely have access to the resources that make their work more productive and ease their heavy workload. Ultimately, it is not just women who are held back, but also their families, their communities and local economies. Rural women have many roles, and they have responsibilities and knowledge that differ from those of men. As farmers, they plant, weed and harvest food crops and tend livestock. As caretakers, they look after children and relatives, prepare meals and manage the home. According to Meinzen-Dick et al. (2014: 56), 'it is important that gender analysis be among the factors of agricultural production. This is important because it will enable researchers and other policymakers to have better insight with regards to agricultural production. Furthermore, it will provide better insights to the constraints faced by both male and female farmers. In addition, it will give data on the different experiences of women and men when studying resource politics. This is important because women have traditionally played a central role in agriculture in developing countries, where on average 40 percent of the agricultural labour force is provided by them.'

In many parts of Asia and Africa, the proportion of the women's agricultural labour force is as high as 50 percent (Shepherd, 2014). With the agricultural labour force being at 50 percent, the study sought to better understand from both male and female farmers what they felt would assist them to become better farmers.

Women as better farmers

The women farmer respondents noted that better access to finance would be very helpful in having them improve the returns from their farms. Others noted that better access to farm inputs such as fertilizers, as well as seeds would contribute to achieving better returns. In addition, support from the government in terms of finding better markets for their crops was also highlighted. 'Better training is needed in ... crop husbandry, financial management, record keeping, mechanization and the provision of financial resources' as indicated by the scheme manager and male respondent. Is important when discussing women's roles in agriculture that certain prisms are taken into account. Doss et al. (2018: 70), for example, stated: that 'the kernel of truth behind the myth that women account for 60–80 percent of food production is important for food security, especially within women's households. This is because women's kitchen gardens or homestead plots are often not counted as agriculture, but they play an important role, especially in providing dietary diversity. In some contexts, women also grow a large share of the staple cereal or root crops that are consumed by the household.'

6.5.3 Women and Land

In the case of the Ahero Irrigation Scheme, many respondents either inherited the land from their grandfathers or their late husbands, within the Scheme. Those farmers who did not fall within these two categories have a working agreement within their family clans: As noted by the following respondents: 'Through land leasing and also my family's share from my father-in-law.' 'the land belongs to the family' and 'Inherited land from my grandfather' as stated by some female and male respondents.

It is for this reason that we need to understand the myth that claims 'women 1 percent or 2 percent of the world's land' (Doss et al., 2018: 71). This is important because women are extensively involved in food production. The data in the Ahero Irrigation Scheme study show us that women do not hold a lot of power with regards to land as land distribution and control are skewed towards men. In addition, the management of Ahero Irrigation Scheme notes that almost all inheritance systems disadvantage women in terms of inheritance and when women legally inherit, they often face strong social pressure to relinquish their inheritance as stated by the scheme manager. The manager asserts that when a woman is mentioned as a successor, towards a plot of land in the scheme, they encounter difficulty with regard to taking ownership. This is because the in-laws view the land as family land and not individual land. They therefore use wife inheritance as a means to secure the land. If not the in-laws may fight the women, to secure the land for their family.

Rural women face insecurity when it comes to land tenure (Samandari, 2017). This was because land accessibility is based on their relationship with men. This situation has not changed over time. This situation needs to change. African customs play a great role when it comes to stewardship of common property resources, and it is for this reason that African women suffer. This is because these customs are made under patriarchal conditions to ensure that men controlled community assets. While this system of land ownership may have beneficial intentions for the community or clan, it impacted detrimentally in ensuring food security to men and women farmers.

It is therefore important that laws be introduced and enforced to ensure that both genders reap the benefits of the land they work on. Women must understand the legal pluralisms of the law so that they can use it to better understand and recognize different situations that may arise about land. Data on land ownership in Africa is scarce. This scarcity is attributed to the lack of accurate data collection and storage. Thus, 'it is difficult to give accurate data with regards to gender ownership of land in Kenya and sub-Saharan Africa as a whole' (Doss et al., 2018: 72).

With regard to the Ahero Irrigation Scheme, a majority of the respondents do not hold titles to the land as the land is trust land under the ownership of the government of Kenya, since independence. Thus, in real sense land ownership in the Ahero Irrigation Scheme area does not belong to any one gender. However, the leases within the scheme are in the names of men and not women. According to ecofeminism prisms, where land titles exist, difficulties exist as there is a lack of documentation with the names of female farmers as landowners. This therefore hinders the chances of women acquiring agricultural resources, such as capital, and other farm inputs that can boost their productivity on the land. It is therefore important that we have policies in place that can ensure that women have equal access and ownership to land as men. In addition, there should be initiatives that can support women farmers to boost their agricultural productivity.

Land Acquisition and Customary Rights

Respondents indicated that women encountered challenges with land acquisition within the scheme. This situation parallels the literature on women's agriculture about the challenges women faced in terms of land acquisition. This is because culture plays a strong part in one's status of ownership to land. In addition, these relationships were encompassed in different ways in which women related with the entire family or their husband's families as a first or second wife, or with regards to their marital status within their community, or the size of the land that was in dispute. However, some did not view their acquisition of land as challenging as they faced no challenges in acquiring land. They nevertheless noted that due to the small size (4 ha) of the land acquired, it was hard to divide the land among their children in inheritance. Respondents asserted that in some cases customary rights were used to gain or claim access to land. The customary practice referred hereto as to wife inheritance. Although, not always practised.

Customs within the Luo community have an integral part to play with regard to different relationships within the Ahero Irrigation Scheme. For example, brothers-in-law are required to 'inherit' their widowed women to allow for the latter's land retainer ship by the kin. Women farmer respondents noted that African culture and traditions run deep within the community in the Ahero Irrigation Scheme and that changing the perceptions within the community will be challenging on matters relating to women farmers owning land. However, the management of the Ahero

Irrigation Scheme has adopted measures to assist women in fighting for their rights when customary rights tend to undermine women's acquisition of land, The Government Land Act (Cap.280) provides that compensation is compulsory when the community land or private land is acquired and is payable to the spouse/s of affected persons as well as to "any person occupying the land and the spouse or spouses of such person" (LA s. 107). The CLA stipulates that there must be "equal treatment of applications for women and men" (CLA, s. 14 (4) (c) (i). Women marrying into a community may not be excluded as members, and their rights to land remain the same unless they divorce and remarry elsewhere (CLA s. 30 (5). The above thereby indicates that there are sufficient grounds for women to appeal against injustices they may have suffered at the hands of the community land committee in matters related to land dealings.

Land Inheritance

As land is a critical asset in farming the study sought to understand, how respondents within the scheme would transfer ownership of this vital asset. Eighty per cent of the respondents noted that they would nominate their sons as their successors on the land title. One respondent was different and stated that he had only a daughter, and therefore he would leave the land to her. The female respondents asserted that they would give land to their first-born sons and live with their daughters. Other women farmers said they would divide the land equally among their children, while others said their children should work hard and buy their land. Yet another group responded that the land would be divided equally among the children, regardless of gender. For some female respondents, the matter did not concern them and was their husband's duty.

Lawry et al. (2017) asserted that many farmers in developing countries hold customary rights as a highly secure measure in the context of local social arrangements. However, this may not be the case regarding the legal status of the country's statutory property regimes. Thus, in such situations, land assigned under customary arrangements was most often statutorily categorized as public land and subject to the stewardship and administration of public agencies. They argued that in some instances where commercial investments took place, the customary tenure arrangements that delivered secure tenure rights to generations of farming families have been overridden and farming families have faced displacement.

6.5.4 *Women and Water*

Women and Water Access

'Water is a key asset in farming and is the source of life. Irrigation is a vital component of meeting the food security needs of the world's increasing population' (Nakawuka et al., 2017: 2). In addition, it is a valuable tool that can increase food production and move small-scale farmers to large-scale production. This is important, especially in areas where erratic rainfall is experienced. This is because irrigation allows for more than one crop in a year to be grown, and can thus help farmers grow high-value cash crops. A manager at the scheme indicated that access to water for farming is gained by pumping water from River Nyando and that farmers pay a fee of Kshs.3,100 ($31) as operational maintenance to have access to the water. All farmers within the 12 blocks of the scheme gain access to water after the cropping plan is agreed upon.

> 'Yes, we do our abstraction from the river; our river is River Nyando, that is our source of irrigation water for the scheme. We pump the water into our system, that is, once the water is in the main canal and then it flows through the canals to the ... 12 blocks. We don't open the 12 blocks at the same time. Once we have opened and there is water in the main system... we see the water to the block where production is done. Then within a block, we have feeders, so once water comes into the second block, then it is opened through the irrigation water users association leaders at the level, so the uptake has a leader who opens to different feeders within that block, we open feeders, within the specific acres within that block, so that is how farmers get access to water....'"... Yes for the main crop we charge operational maintenance Kshs3,100, for any other crop if you want support in terms of irrigation water you are expected to pay a small maintenance levy or they pump on their own, so what they do they get water through pumping , for a small levy...'

Challenges of Water Access

Ninety per cent of the female respondents mentioned that they did not face any challenges in gaining access to water for domestic purposes. In regards to water for agricultural use, they assert that they do encounter some difficulties. Some of the female respondents asserted that *'Yes, in case of power blackout, which at times goes on for one week, there is always a lack of water since the pump uses electricity. This is a major challenge and there is nothing we can do until the power is back.' For men on the 'They are no other challenges apart from time; water takes time to reach your farm but we are just patient.'* And *'As for me there are no challenges in getting water' 'Yes, if there is power, we are okay. If there is drought, we have a lot of challenges? It also depends on how many acres you have?'*

Nakawuka et al. (2017: 3) assert that 'smallholder irrigators normally suffer from economic water scarcity. Water resources might be available in a community, but individual smallholder farmers might lack resources that allow them to access and use these water sources adequately.' In such cases, the need arises to work in groups (e.g.) farmer cooperatives, or to take up credits or be dependent on government or donor contributions. Thus, making water accessible to smallholder irrigators has been suggested as having the greatest potential of achieving food security in sub-Saharan Africa. It is important to note that some of the female respondents felt that they had to fight to gain access to the water. *'In case there is no power we really suffer since the pump uses electricity', 'Yes, most of the time you find water being directed to farms at night. So, as a woman that time is difficult to access water to the farm. So, it will force you to hire someone to do that for you.'*This is because water becomes scarce during drought periods, and when the Irrigation Scheme lacks power to pump water to the various farming blocks within the scheme. Water accessibility, is important in ensuring crop productivity levels for small-scale farmers is productive. The control over natural resources is limited for the majority of women within Ahero Irrigation Scheme. This limited participation of women in decision-making has the potential to fail government policies and programmes according to (Le, 2018). Magill and Benedict (2019) assert that the value system for water has become the key issue about who has access to water and at what cost. This can be seen from the comments of the respondents. This is because once water becomes a scarce resource it needs to be commodified and priced according to the market; thus making water become one of the many 'things' a consumer can buy. Water as a natural resource should not be bought, yet the need for water in nature is quite compelling for human beings. This is because marginalized communities cannot afford to pay the market price for water when it is privatized. Thus, 'power, poverty and equality' factors can be seen in the commodification of water. Hence, the reason Ahero Irrigation Scheme charges farmers fees to maintain water access points. It, therefore, becomes important for ecofeminism to highlight the important relationship that gender has with nature, and most importantly why this relationship should not be ignored. As one respondent stated 'Everyone can just access water without any fear or favour from anyone.'

Respondents noted that they did not face any cultural barriers in accessing water. One respondent stated that the barrier was mostly organizational, and not cultural. She stated that there is 'no cultural but organizational challenge that only elected people should be responsible for the distribution of water to the fields'. According to Nakawuka et al. (2017: 7), 'aside from infrastructural challenges, there are institutional challenges such as the weak functioning of water users' associations (WUAs) which tend to lead to ineffective control systems, inadequate monitoring and ineffective enforcement of activities around the scheme.' Nakawuka argued that water management in small irrigation schemes is usually inefficient. In Ahero Irrigation Scheme, respondents assert that one of the challenges they face in regards to accessing water during drought periods happens when there is no power to pump water from the River Nyando to the 12 blocks within the scheme. Farmers whose plots are upstream in the scheme receive ample water; unlike the downstream farmers who especially under conditions of reduced water supply often do not have water reaching their plots. This, we have noted, plays out in the irrigation scheme: 'Yes, most of the time you find water being directed to farms at night.' So, as a woman that time is difficult to access water to the farm. So, it will force you to hire someone to do that for you. Respondents also note that WUAs, and the Ahero Irrigation Scheme cooperative are not able to cater to the needs of members due to lack of sufficient funds to run activities within the two organizations. The dominance of masculinity in irrigation management is so strongly normalized that it is not necessary to mention 'men' alongside irrigators, as everybody 'knows' that agricultural water management is men's work Lebel et al. (2019).

Lebel et al. argue that even though women also use water for agricultural purposes, water is still a male resource that is governed by men. In the Ahero Irrigation Scheme, the case holds as stated by one respondent who stated: '...men think they are overall in water management...' Women within the scheme assert that men are in charge of water management and men therefore make all the decisions related to water distribution within the blocks.

Water Recipient Time

Unlike some other cultures where women face cultural barriers in accessing water, this was not the case in Ahero Irrigation Scheme. The scheme made provisions for clean water for the farmers within the scheme. However, the distance from these water points varied and was dependent on where the water point was situated. In some cases, respondents noted that a water truck with clean water was sent to the various villages within the scheme to give them access to clean water, or others access it from the headquarters of the National Irrigation Board which for some may be their nearest watering point. According to Van Koppen (2016: 5), 'access to water implicitly refers to water in the right quantity and quality, at the right moment, and in the right place.' In short, water is available for use. To use water and derive the benefits from it, people need water in the appropriate quantity and quality, and at the appropriate site. Thus, they gain access by moving to the water, for example, through permanent settlement patterns, or by long-term pastoralist transhumance, or seasonal flood recession farming.

Water accessibility within Ahero Irrigation Scheme is not a problem. This is because the management of the scheme has ensured that water is easily available for domestic purposes. This therefore reduces the amount of time spent by women when collecting water for domestic use. In regards to accessibility for agricultural use, water for this purpose seems to take a much longer period. One male respondent noted 'For domestic is 5 minutes; for farm use it takes approximately 3–4 hours for the water to reach my farm when pumped.' A female respondent 'For domestic use I cannot approximate because I have my tap in the compound; for farm use it takes approximate 6 hours.' Female respondent 'For domestic use—I have a tap within the compound, for farm purpose—the pump is located at approximately 4 km away from my farm, I have to wait for it to be pumped.' 'For irrigation, it is once after 14 days.' As asserted by Van Koppen (2016) water is an important resource, its absence in the right quantity or quality will always force people to look for it or to move near it as asserted by Van Koppen (2017).

6.5.5 *Financial Resources and Assets*

Access to financial resources and assets is a problem for both men and women within the Ahero Irrigation Scheme. Farmers within the scheme had the privilege of accessing funds via a revolving fund where farmers could access loans, during planting seasons. However, due to some challenges, the funds were not easily accessible to all farmers in the scheme. Thus, farmers had to resort to other means to acquire funds to support them during planting seasons. As mentioned by one female respondent these were businessmen outside the scheme who loaned the farmers money and, at most times, turned their harvest into a loss instead of a profit.

> 'If I do not have money, I normally go to the businessmen in the market and they can give me Kshs, 2,000 and then when it comes to the time to harvest, they can come and collect one or two bags of my harvest. ... This is because if you borrowed, for example, Kshs, 2000 from the businessman and he comes to collect the bag during harvest time, he/she can sell the bag for even Kshs.5, 000 so he benefits from you - after all the work you have done. He/she will not collect the Kshs 2,000 that you borrowed; he/she will just take the bag of rice which is now priced at Kshs5,000 - yet he only gave you Kshs 2,000...like last year a bag of rice was selling for 8,200 Kenya shillings...so you end up going at a loss. It's like you are working for someone else...sometimes the office for the revolving fund can assist you if you come to see one of their officers; if you are a good farmer they will assist you and collect a small interest ...if there is no money then they will not be able to assist you. The office has really helped us a lot.'

On the other hand, those who were in good standing with the scheme could buy seeds and fertilizers from the revolving fund as explained by the scheme manager: '...there is a farmer's body that is in charge of production, so they receive inputs, seeds and chemicals...' Others stated that they use their own savings to buy their farm inputs, while others use the seeds from their previous harvests. One male respondents asserted that 'I have 2 ha; I have to lease one to somebody. That money can assist me because ...the scheme does not give you any money to buy fertilizers.' A female respondent asserted 'Personal savings, sale of personal assetsassets' 'I normally keep rice after harvesting to use as seedling for the following planting season, fertilizers I buy from agrovet.'

Women in Ahero Irrigation Scheme, do not have easy access to finance. This is because they do not have the right to access resources (land) that can enable them to access finance. This is because financial institutions that lend money require assets as a form of collateral, and most of the assets such as land are mostly in their husband's name. In addition, most financial institutions will not accept crop harvests as collateral as they have no guarantee that the harvest will be sufficient to cover the loans. Wulandari et al. (2017: 1) argue that 'finance is an important asset in sustaining the production of agricultural commodities. However, access to finance for the rural poor is a major challenge, especially in developing countries, such as Kenya. This is because they lack adequate financial knowledge of the various financial products, and are thus unable to make informed financial decisions.' In Ahero Irrigation Scheme, women within the scheme are not well educated, as earlier stated 67 percent of the respondents, have primary education, with 24 percent having secondary education. Financial decisions are not to be taken lightly, and need some form of information to enable individuals make informed decisions. In addition, financial information is not easily made available in rural areas, and is sometimes written in English jargon, which may not be easy to understand, especially for rural women. It therefore becomes important for women to be educated to certain levels, to enable them learn ways and means in which they can acquire finances that can assist them engage in agriculture. It is for this reason that the next question in the study sought to know how women access financial resources, and from whom do they access these resources.

Women's Access to Financial Resources

Many of the women in the scheme acknowledged that they did receive support from their husbands. Some of the men also asserted that they too received some financial support from their wives. If they did not receive it from either their wife or husband, they received it from their children. This is important to note because according to Doss et al. (2018), female heads of households who receive remittances from a migrant husband maintain social connections with the husband's family and expect to have their husband return. In addition, they face different opportunities and challenges than a widowed female household. Doss et al. (2018: 70) assert that 'this is important in debunking the myth that 70 percent of the world's poor are women because aside from casting women as victims rather than as contributors to food security, it focuses on women as disproportionately poor and makes female-headed households appear as more vulnerable to poverty and can therefore distort the design and implementation of programmes and policies.'

One respondent noted the following comment with regard to receiving support from her husband: 'Yes, my husband always assists me because the process of acquiring financial resources is too long for women for we are not trusted.' This thus showcases the team effort between men and women (or couples) in the scheme, and the impact this has on their farming activities.

Financial Incomes

Two hundred thousand Kenya Shillings Kshs.200,000 ($2,000) was the highest amount received during a harvest. The respondent stated that she obtained these funds from selling her harvest. The second highest amount received was Kshs.150,000 ($1,500). A conclusive amount cannot be stated in this section, as respondents did not state the time period in which these funds were received. In addition, the study cannot make assumptions with regard to the income received from harvest as this was not tracked for a specific period. It is important to note that some women within the scheme do receive access to funds through the revolving fund and harvest returns. It is also important to understand how the women receive access to their returns from the farm. For example, one female respondent stated:

> 'I have received around 60,000 to 80,000Kenya shillings ... My husband ... takes and divides the money. He gives me a little and he takes a good chunk. Women here do not have accounts in their names. We are paid from the irrigation board here to the bank account, or you come here and collect the money, or you come and collect cash from the revolving fund because you have no account. He (i.e., the husband) comes and collects the money and pockets it all - the hard cash. If he gives you Kshs 10,000 you cannot complain ... because if you complain... , you can end up being beaten ...he also becomes harsh and at that time he can tell you to pack your things and go ...because he has money ... when the money is all finished, then he will remember he has a wife.'

We must understand that not all women will receive funds, from the selling of their crop harvest. This is because it is left up to the man to decide whether he wants to share the funds or not. In other cases, as explained by respondent one female respondent, the man may decide not to share the funds at all. It is therefore important that women within the scheme are empowered financially. This will strengthen them financially and empower them to make financial decisions.

Women's Financial Challenges

When it comes to accessing financial resources, within Ahero Irrigation Scheme, women within the scheme face numerous challenges. As some of the female respondents asserted *'We lack items to provide security for loans, and our husbands think we misuse the money'*. *'Women are not normally regarded as real farmers, hence it becomes difficult for them to access financial resources.' 'the assumption that women are poor loan payers–hence, high rates of interest accrue during repayment period.'*

On the other hand, some men did not feel the same as women, when it came to this issue: *'Lack of knowledge about financial resources'. 'Poor repayment by women and misspending the money.' 'Risk taking; misuse of funds by some women, thereafter some women obtain loans. Suppose you secure loans from the women enterprise fund and fail to repay on time, an auction will follow.'* Then, again some women in the scheme felt that women were not good with financial resources *'Many people think that women misuse financial resources by regarding them as their own, and they use them for their own budget.' 'Misuse of available funds in buying luxuries'.*

The respondents felt that women mishandled money; others felt that women lacked enough information about how to obtain loans, while others felt that revolving fund defaulters were to blame. It was important to note that some of the women respondents felt that discrimination still played a part in their failure to access financial resources. The application process was also noted as being a hindrance to gaining access to funds. High interest rates were also mentioned as part of the challenge in seeking loans. 'Yes. High interest rates from money lenders. I believe that when the interest rates are favourable, I can get access to funding.' 'Lack of skills—if women are trained on how to use some of the resources it will help a lot.'

> 'Yes, gaining financial resources requires a lot of procedures to be followed. You need to be patient and determined to get access.' 'Yes, male partners ask for several favours before they can assist you to obtain financial assets. We need the women enterprise fund to be devolved to us down here.'

When farmers are subjected to challenges, it is interesting to analyze the financial choices they make to enable them to meet their agricultural investments and spending (Omobitan, Khanal and Honey (2019). Thus, given these circumstances, informal credit sources are unquestionably the most popular source of finance to the rural and urban population because the formal credit sources have scared many food crop farmers due to the encumbrances. The informal financial sector is an unorganized sector that consists of money lenders, relatives, friends, neighbours, landlords, traders and groups of individuals who operate mainly in the rural setting, and who seldom engage in lending money (Gbigbi, 2019). These informal financial groups are the major providers of funds for the promotion and development of small-scale farmers in the rural areas. Furthermore, they play a key role in resource mobilization and allocation in developing economies. Historically, financial services interventions for smallholders have tended to focus on loans for working capital to access non-labour inputs (such as seeds, fertilizers and pesticides) with some interventions aimed at longer-term, community-level projects such as irrigation or processing. With overall credit access being low, it's hard to tell whether men and women have different credit needs in agriculture. This is because both genders require credit when engaged in agriculture. In the case of Ahero Irrigation Scheme, both genders have financial needs, but women need to access credit more but are unable to do so due to loan requirements. Furthermore, men are more likely to take decisions about large investments, thus it seems likely that men have a higher demand for larger, longer-term loans for irrigation or mechanization. It is therefore important to note that farming households have a definite demand for working capital loans. However, these working capital loans are currently being met mostly in the informal sector.

The major obstacle to effective rural banking relates to the lack of banking facilities, which facilitates the patronage of moneylenders (Gbigbi, 2019). It is claimed that in most rural areas where banking facilities are deficient, informal credit and savings markets are predominant and that farmers patronize moneylenders for both consumption and investment credit during land preparation and planting season. Inadequate access to productive resources limits women's livelihood options and exacerbates financial strain on women. By improving women's access to productive resources, their potential could be unlocked, thereby enhancing their efficiency and productivity. Therefore, the status of women farmers in accessing resources in different regions is needed to assess and design suitable strategies that will improve their access to resources (Sarkar et al., 2018). In regards to the Ahero Irrigation Scheme, the implementation of women's farmers groups may be key and the first step in tackling access to financial resources. This is because the groups can be used to educate women on financial resources, and the same group can obtain assets and help members access credit that will enable members to purchase agricultural inputs that they may be lacking. These same groups, in addition, can be used to help better train and equip women with new farming techniques that will help them increase crop productivity.

6.5.6 *Technology and Education*

Many of the respondents within the scheme assert that tractors are used in tilling their farmland. In addition, the use of tractors has reduced the time taken in tilling their farmland, from 2 to 3 days to 2 to 3 hours. This is important as it shows how mechanized farming quickly impacts farming productivity, as brings about efficiency and reduces time spent tilling the land, thus creating time for other chores to be done within households and agricultural farms. The male respondents noted *'We are using the tractor … so the tractor comes and tills. It comes from the irrigation scheme - from the co-operative society which was formed by the irrigation scheme'. We use tractors from the co-operative society and they provide their operators. It can take upto 5 hours to till all the land.' The female respondents noted 'I till my farmland using a tractor which takes 20–30 minutes.' 'Co-operative tractors; it takes 4- 6 days for the tractors to reach your farm after application, and it can take 2–3 hours.'*

'Labour-saving technologies will have differentiated impacts on women and men as long as tasks in agriculture are gendered' (Slavchevska et al., 2016: 25). This is because some tasks are viewed as women's responsibility and others as men's responsibility. Hence, some technological innovations may displace female or male labour, and thus be resisted by both, even though they have the potential to improve productivity. In the Ahero Irrigation Scheme, this is not the case with regard to incorporating mechanization into the scheme. A majority of the farmers when asked what kind of equipment they used when tilling their farmlands said that tractors were mostly used. *'We use tractors from the co-operative society and they are provided with their operators. It can take up to 2 hours to till all the land'*. However, this was not the case for all the farmers, as others stated that they still used ox- drawn ploughs to till their lands, as well as hoes and other traditional tools to farm. This perspective was best summarized in the words of one male respondent who stated, 'Oxen plough, it takes 2–3 days.' A female respondent stated that 'Labourers who take 2 to 3 days tilling one ha of land.' In addition, some of the farmers were trained to use the equipment while in school, while others said that they trained themselves to use the equipment. A majority of the farmers stated that the tractors that till their lands came with an already trained driver who knew how to operate the tractor, as well as the other machinery that came along with the tractor. Farmers in the scheme also used traditional farming tools as stated by a female respondent who stated, '... '*Jembe*', '*Pangas*', tractor, rakes,... I learnt how to use them through knowledge that I gained in school.' Another Respondent stated, '*Jembe*'- I was taught how to use it by an agricultural engineer in the nearby irrigation scheme.' It was also noted that the time taken to till the lands varied. This came about as a result of the size of land one had and the tools they used. As one female respondent stated. 'Co-operative tractors it takes 4–6 days for the tractors to reach your farm after application, and it can take 2–3 hours.' Hence, 'for newly introduced practices and technologies to be adopted by farmers, a lot of time must be invested in farmer awareness' Nakawuka et al. (2017: 8). In summary, it came down to the resources that one had at their disposal, as other farmers may not have had access to the same resources as others, even within the same scheme. As stated by a female respondent, 'Oxen- plough it takes 2–3 days.'

'New technological innovations tend to benefit men more than women, thus, lessening the workload of the former and increasing the activities linked to women, such as transplanting, weeding, harvesting and processing' (Chauhan, 2015: 76). Furthermore, it has been argued that modern biotechnology benefits men more than women. Also, Saugress (2002), and Bryant and Pini (2006) argued that the aim of using machinery and agricultural technology was to remove women from the agricultural sector through patriarchal ideology. Saugress (2002), and Bryant and Pini (2006) claimed for instance, that 'a tractor is a symbol of men's power and is a means of dominating women'.

Men tend to prefer irrigating cash crops such as tomatoes, while women favour leafy vegetables, which can be sold over a longer period and allow them to retain control over the income. Additionally, women in pump-owning households have less influence over production decisions but still use the pump for their own plots. Given these dynamics, technology adoption should not be viewed as an end goal but rather as a means to understand its varying impacts within households and to align technology promotion efforts with specific development objectives (Lefore et al., 2017a).

'It is a government institution; we normally have farmers capacity building at least quarterly ....We have not had much, apart from what we received from the central government, but there was one that was done in 2013 when the World Bank had a project and supplied farmers with inputs and when they are coming up with new products they give training other independent organizations No (training)'.Training is a very important tool in farming; we are informed that farmers in the scheme undergo regular training, to learn new farming methods. This is important to help farmers gather good farming practices as well as increase their crop harvest' as explained by the scheme manager.

Farmer Training

Farmers within the Ahero Irrigation scheme indicated that they had received training from the Kenya Seed Company, and a local non-governmental agency. This training enabled them to gain new farming skills as stated by a female respondent who stated that '...gained new farming skills...improved the production of rice.' Another respondent stated that; '...the training helped me learn how to plant rice in rows.' This improved my yield.

Training opportunities within Ahero Irrigation Scheme are beneficial, as farmers within the scheme benefit, from new farming methods taught, as well as learn how to use new farm inputs, that increase crop yields as stated by several respondents. '*We have not had much training, apart from what we received from the central government. There was one set of training that was done in 2013 when the World Bank had a project and supplied farmers with inputs, and when they came up with new products they provided training. Other independent organizations–no, they did not provide any training' 'Yes, we do, but rarely. The last time we received the training was about 6 months ago*' Seventy per cent of farmers stated that they received training and that the training helped to improve their yields, as well as their knowledge.

However, others stated that they had not received any training at all. For example, one female respondent stated: 'Yes-Received nice training on improving yields and soil.' While another stated, 'Yes, Row planting has increased the yield.' Sulaiman (2018) asserted that recent analyses of curricula and materials revealed that despite visible guidelines and directions, the status quo reigns supreme. Namonje-Kapembwa and Chapoto (2016: 2) argued that 'it is imperative to design technologies that fit men and women if both genders face constraints in their rate of adoption of technologies. However, if the difference in the rate of adoption is because men and women face different constraints, then it is important to address this unequal access.' This is because it is hoped that such actions will promote agricultural growth and will increase women's participation and contribution in farming and income generation. There is a need to invest in higher agricultural education to train large numbers of personnel who can modify and adapt applied research and then extend the knowledge to farmers (Nakawuka et al., 2017). These extension services are a source of knowledge on various agricultural packages to farmers. He asserted that the four East African countries (Kenya, Uganda, Tanzania and Rwanda) lacked qualified extension workers and that the number of extension staff needed to reach farmers was not enough because they lacked incentives to retain them.

'In Kenya for example, the extension staff to farmers' ratio is 1:1,500 and the extension officers are limited by lack of fuel for the motor bikes to reach farmers' (Nakawuka et al. 2017: 10). In addition, project developers may offer technical support in operation and maintenance of irrigation schemes at the beginning of the project but later leave the management and maintenance to the beneficiary farmers. However, many farmers expect that they will always be provided with this support and are not willing to contribute to such activities from their own time and resources due to a lack of ownership. In regards to the Ahero Irrigation Scheme, farmers within the scheme have in the past been reluctant to take ownership, of certain tools in the scheme, due to costs of maintenance and lack of training. Kirui (2019: 4) asserts 'the importance of education and literacy in enabling farmer's access to information and knowledge regarding production technologies and market opportunities as it allows them to make better decisions.' He claims that studies have shown that farmers with just a few years of basic schooling are more likely to adopt and correctly apply agricultural innovations.

Applying this framework to technology adoption allows for a better understanding of how gender impacts smallholder technology adoption in households. The distribution of rights to technology shapes whether technology adoption is sustained or adopted given that the anticipated distribution is likely to influence men's and women's interest in trying out the new technology. However, the lack of education, according to the respondents, does support this argument '*A woman who does not have any education, she does not know even how to sell rice. Therefore she cannot grow good rice, she cannot manage to complete planting in good time - so she cannot manage to get something.*' '*Yes, some women lack knowledge; they do not believe in new techniques.*' Government plays an integral role in the governance of resources. In addition, politicians and political will play an important part in ensuring that these resources (land, and water) are available.

6.5.7 *Agricultural Politics*

Politics is entwined within every fabric of society. From the smallest rural community to the greatest nation, there is a group of individuals who are known as the political ruling class, who are generally empowered by their people to rule and guide them. In Africa, as Poulton (2012: 2) argued, 'Africa's political elites have been successful in suppressing any genuine pro-farmer political movements.'

In the Ahero Irrigation Scheme, this tone was no different: as stated by the scheme manager, *'Politics came in and some leaders were telling farmers this is your money why should you return it. It was given as a grant and the farm is yours, you are the one who is benefitting, you do not need to recover, just take everything the government has given, so because of that the project did not continue because there were no funds.'* 'The primary way of gaining the allegiance of a group has been to give it access to government resources by granting its leader(s) control of particular state institutions.' (Poulton & Kanyinga, 2014: 5) Thus as stated by one male '*Yes, political affiliations affect the distribution of resources regionally*'.

The political process … is one of perpetual coalition building—of the formation and reformation of coalitions of local organizations that have no common interests other than the control of the state and its resources. Stated simply, 'the objective of politics is to control the state for the purpose of funnelling state resources back to one's supporters and local communities. Politics is local and patronage is usually king' Poulton and Kanyinga (2014: 5). As stated by one female respondent, '*… Politics has gone tribal; the government does not help us*'.

The Ahero Irrigation site is located in the western region of the country. This region is viewed as a region that often supports the opposition. Thus, the residents within the scheme felt that they are being punished for voting for the opposition. For example, one female respondent stated: 'Yes, our scheme is in the opposition area so the government does not supply us with equipment,' while another stated, 'The government has forgotten about the farmers.' Poulton (2012: 5) asserts that 'major agricultural value chains have a strong association with particular groups. This is because major ethnic groups live predominantly in one area of the country, in addition to the major cities. This has made state-supported agricultural institutions powerful instruments of patronage and central features of Kenya's agricultural policies.' About local politics, some of the respondents felt that the local leaders were in office to satisfy their own selfish and personal interests, while others felt this was not a concern and responded 'No' to this question. For example, one female respondent, stated: *'it does not affect me'*, while another female respondent, stated that: *'Yes, some leaders satisfy their personal interest.'*

The farmers felt that the County Government of Kisumu has been beneficial to them. This is because the County Government was able to purchase tractors for the farmer's cooperative, which they used to till their lands. Hence, they felt they had benefited from the County Government. Some farmers stated they received bags of fertilizers from the National government, which helped them to save some money. *'We received tractors from the former governor of Kisumu Ranguma; he stood as a guarantor for us so that we could receive tractors, he saw the way we were suffering and felt that by giving us the tractors it would be a good way to help us. ... He used the county to guarantee the tractors to the farmers.' 'Yes, I got fertilizer - 3 bags from the county government and I saved some funds for other use'* However, the scheme also received funds from the National government to train farmers on a yearly basis. These training sessions were important in training farmers in new farming methods and enabled them to produce greater output from their farms. On the other hand, there was a big gap between the National government and the County Government in the coordination of projects and in assisting farmers in the scheme, and in meeting and understanding both governments' objectives as well as their own. *'Yes, the county government took some farmers to Uganda on the farmers' exchange programme.'*

Agricultural Policies

It is interesting to note that the farmers within the scheme benefitted from a few initiatives during the last 5 years, and most significantly from the County government of Kisumu. The County guaranteed itself to the Japanese government and was able to purchase tractors for the farmers that they used through their cooperative society. For example, one male respondent stated: 'Yes, we received tractors from the county government almost two years ago.' In addition, the Government's economic stimulus package—that is, referred to as the revolving fund—also played a key part in assisting some farmers within the scheme, while others noted that they have not benefited from these schemes at all. One female respondent stated, 'No, I have not benefitted from either of them.'

It is also important that many of the respondents noted that they had very little interaction with people in government with regard to developing or creating policies that would support them within the scheme. However, the farmers indicated that some discussions were held but these were not at an in-depth level. For example, one female respondent stated: *'Yes, but they rarely fulfill their promises'*. However, we were informed that when need be, the government representatives come together in consultative meetings to tackle the issues at hand. As stated by one of the scheme managers: *'When there was a new act,they brought it down for farmers input and when they did that, the local MPs and governor came together for a consultative meeting, then it was discussed in parliament.'*

Government Actions to Improve Women's Role in Irrigated Agriculture

The farmers acknowledged that women were important players within the scheme and that the national government should consider the following key elements in improving women's roles in agriculture in Kenya. These included: access to finance is a vital part of any sector and gaining access to funds will better assist women within the farming sector. In addition, better access to farm inputs such as seeds and fertilizers at subsidized rates is also viewed as vital. Educating and training women on better irrigated farming methods is also vital not only for them but for all farmers within the irrigation scheme.

This would enable the irrigation board improve farmers yields, manage their crops and enable them to have a better understanding on how to manage resources. The provision of markets where farmers can sell the crops was also important, and the elimination of middlemen from the equation will better increase the returns farmers receive and reduce exploitation (by the middlemen). As one respondent stated: 'eliminate middlemen from the market, provide more training for farmers and provide subsidies to farm inputs.' One of the respondents noted the importance of educating the public on the importance of having women inherit land. A key element in helping reduce the stress of women, especially widows, was to have some form of security for their families, especially regarding land tenure. This should not only be in terms of financial security but food security as well. As one male stated: 'The government should protect women in their inheritance farms… to reduce discrimination.'

Women in Ahero Irrigation Scheme, play an integral part in irrigation farming. Men in Ahero Irrigation Scheme acknowledge the important role women play in farming and irrigation. The management of Ahero Irrigation Scheme, acknowledges this as well, and have measures in place to support women in this role. This is a start, and more needs to be done to support these efforts. The Government of Kenya can support these initiatives further, by developing policies that support women in agriculture. Numerous respondents mentioned ways in which the government of Kenya can support them.

One female respondent,

> 'Actively involving women in agriculture production, making it easier for women to access loans, modernizing agriculture and doing away with traditional beliefs affecting women.'

> 'The government should protect women on farms they inherited… to reduce discrimination.'

> 'The national government should make it easy for women to own the land. The local/national government should clearly support the women farmers in a way that it reduces labour.'
> 'Women should be provided with funds to reduce borrowing. The government should provide water for irrigation.'

It therefore becomes important for the Kenyan government and other stake holders in the agricultural sector to develop policies and initiatives that will support women in the irrigated agricultural sector, and help them grow in the sector.

6.6 Conclusion

In the many discussions and documentation noted in the study, it was acknowledged that women played a vital role in irrigated agriculture within the Ahero Irrigation Scheme. The scheme management understood this role and knew the vital role women played when it came to planting, weeding and harvesting of the crop. However, some of the women within the scheme did not seem to fully comprehend their role within this context and the importance attached to it. It was within this context that some men within the scheme acknowledged the vital role that women played, while some of them did not fully acknowledge this. The gender gap within irrigated agriculture within the scheme seemed not to be so prevalent. This was because both males and females took on different responsibilities within the scheme and within the agricultural context of rice farming, at the different stages of the crop cycle. The management of Ahero Irrigation Scheme acknowledged the contribution of women farmers within the scheme. However, more needed to be done to further support them in terms of agricultural training programmes, finance, crop husbandry and access to financial resources and assets. Although the scheme management played an important role in fighting for women about land ownership status within the scheme, more needed to be done with regard to setting up policies that incorporated socio-cultural factors that hindered and prevented women's access to farmland, including family disputes. Vital data with regards to the challenges women farmers faced within the scheme was able to be gathered. Many farmers noted the need for financial assistance, training on proper farming methods and better access to farm seeds and fertilizers, as well as better access to land. This was because some cultural aspects hindered some women from inheriting land from their husbands due to cultural land policies which differed within the various 12 blocks in the scheme. Credit facilities were available to some farmers within the scheme through the revolving fund but were not accessed by all farmers. The farmers, both male and female, acknowledged the need for better and more frequent farm training sessions within the scheme to better equip them and train them on new irrigated farming technologies. This, for them, would have increased their own knowledge and crop yields. The National Irrigation Board should look into creating policies that promote gender equity about farming within the scheme. Equity, rather than equality because social-cultural aspects did not require one gender, but both genders so as to initiate changes that can be widely accepted by all within the various communities in the scheme. Politics also had an impact on the scheme. Some farmers referred to the Mwea irrigation scheme, which many saw as a model of what Ahero Irrigation Scheme should be. However, to them, Ahero was not a model irrigation scheme and lagged in terms of marketing and the sale of rice to other counties. The Kenyan Government therefore needed to liaise with many stakeholders in the irrigated agricultural sector to see how they could improve farmers' accessibility to farm inputs, such as seeds and fertilizers. Furthermore, this should not be at the national level but should be a collaborative effort between the national government and the county government, with the latter closer to the people. In addition, this will help create policies that positively affect those on the ground. This should be a bottom-up approach, instead of a top-down approach, to have a greater impact on the small-scale farmer. In as much as the study sought to better understand women farmers in irrigation, the study gained insight into the challenges farmers face in irrigation farming. Hence, there is a need to develop policies that are gender inclusive, as well as work with communities in developing social-cultural means that allow better access to land, finance and training which hinder smallholder irrigation farming. Lastly, projects that require funds to be disbursed to farmers at local levels need to be properly communicated through all channels. This is to prevent many within the different stakeholder portfolios from having different views and agendas. This then reduces the risk of having mixed messages sent to farmers who are the final recipients of the funds. This would help them take ownership of the funds and better manage the resources given to them to improve their farming skills as well as their welfare.

Women in the irrigated sector in Kenya, particularly within the Ahero Irrigation Scheme, face numerous challenges as farmers. These challenges are similar to those faced by women in India, Nepal and Bangladesh, where women are often regarded as secondary to men in farming. Additionally, they lack control over critical natural resources such as land and water. Despite these obstacles, women in the Ahero Irrigation Scheme have demonstrated resilience by securing support from the Irrigation Board, men, and the broader community. This support is evident in the formation of committees within the irrigation scheme that address land disputes. The concerns raised by these women highlight that they are not just farmers but also stewards of water, recognizing it as an essential resource for both farming and household use.

The resilience of women in the Ahero Irrigation Scheme symbolizes the broader struggle between nature and human intervention. Just as nature resists destruction from human activities such as climate change, women fight to sustain their families within their ecosystems. The concepts of nature and nurture are deeply intertwined with women, much like nature itself embodies nurturing. In the rice fields of the Ahero Irrigation Scheme, these women showcase their expertise in irrigated farming, tirelessly working to nurture their crops and provide for their families, regardless of their educational backgrounds. Their primary goal is to ensure their families have food.

The power of women in the Ahero Irrigation Scheme lies in their determination to continue farming on the land they have inherited from their husbands and their advocacy for policies that encourage more women to participate in agriculture. Their strength is reflected in their ability to overcome challenges and their resilience in working with nature to produce food for their families.

CHAPTER 7

Agriculture (4.0) Women and the Future of Farming

7.1 Introduction

Agricultural systems worldwide have continuously evolved over time. The Agricultural Revolution, begun when human beings transitioned from traditional methods of growing foods to living on farms. This was followed by the Second Agricultural Revolution, also known as the Agrarian Revolution in the eighteenth century, which significantly grew due to the use of machines and increased the output levels of both the land and crops. The third Revolution happened in the 1960s in Asia, and was called the Asian Green Revolution. This revolution introduced a technology-driven approach that included improved seed varieties, irrigation and the use of inorganic fertilizers. Today, agriculture is transitioning to a new evolution that is called Agriculture 4.0. A universally accepted definition, of this new revolution is yet to be identified, but the new evolution represents a shift where farmers are adopting and integrating new technologies that include the use of computers and drones and robots to shape the future of farming. This revolution is driven by the principle of sustainable intensification, focusing on sustainable food production and the development of more efficient agricultural systems (Jellason et al., 2021). The proposed solutions challenge the conventional association between digitalization and food production. The digitalization of farming systems is often framed as 'inevitable' and primarily driven

M. O. Jimoh and S. Narsiah, *Unveiling the Power of Women in Irrigation*, https://doi.org/10.1007/978-3-031-94174-0_7

by the need to meet the food demand for the worlds growing population. In the 1990s The Asian Green Revolution, ended having alleviated food crises, reduced poverty, and offered valuable lessons for sub-Saharan African (SSA) countries (Rose et al., 2021). However, it remains controversial. Critics argue that smallholder farmers struggled to compete with larger farms, exacerbating inequality within the agricultural sector. Additionally, the increased use of fertilizers contributed to the eutrophication of streams and lakes (Jellason et al., 2021). In 2006, AGRA (the Alliance for a Green Revolution in Africa) was founded by African agricultural policymakers, with support from donors such as the Rockefeller Foundation, UK Aid, USAID, the Bill and Melinda Gates Foundation, MasterCard Foundation. The organizational establishment was inspired by the Asian Green Revolution. The main aim of AGRA was to drive Africa's agricultural transformation by promoting improved seeds and fertilizers. Most importantly, AGRA's goal was to build smallholder farmers with agricultural skills and practices that would increase their crop yields. However, evidence suggests that AGRA has not met this ambitious target. Critics argue that its shortcomings stem from the use of Western style farming methodologies to farm in the diverse SSA system, high cost of inputs that fail to translate into high yields and over reliance on one type of farming. This approach has led to reduced crop and dietary diversity, raising concerns about its long-term sustainability (Jellason et al., 2021). In addition, the social impact of this revolution has been called out in recent papers. Furthermore, how will the distribution of these new technologies be shared out among farmers on the continent. This is because in the context of collaboration the importance of social sustainability is a significant shortcoming and suggests that the new agricultural age (or 'agriculture 4.0') should be driven by the core concepts of sustainability that benefits people, the planet and production and is holistic not one sided. In the next section we draw upon SI (sustainable intensification). The key principles within this are people-social, production-food, planet-environment from a people perspective, primarily women (Rose et al., 2021).

7.2 Challenges for Agriculture 4.0

Governments today are gradually using technology to address issues such as environmental degradation and hardship. Agricultural technology is now touted as the new idea to produce more food for the world's growing population. The lack of enough food being produced is being questioned on the one hand, because food insecurity is caused when people do not have access to the food in the first place (Rose et al., 2021). Many see the promotion of technology as an easy solution to the problem. However, this strategy can look simple to high profiled individuals who seek to redirect the focus from the social inequalities. However, the skewed allocation of food caused by economic disparities and gender among other forms can also be a major cause of food insecurity in both the Global North and South. One can therefore be seduced to believe that the use of technology is a simple way of solving problems. This is because if not well developed the technology will fail to achieve the sustainable benefits to people. Rose notes that there is some potential controversy with the fourth revolution, which will change farming systems.

7.3 Land Access

Land is key to the production of food and its accessibility, in addition, it is one of the assets that is important within the food ecosystem. Cultural norms depict access to land through many points. However, we note that in the Global South accessibility for women is different as it is subject to social norms and culture. For example, land in Muslim culture is permitted to be inherited by both males and females. However, most females transfer their land estate to their patrilineal relatives. In sub-Saharan Africa, land is androcentric, and is mostly given through the masculine line. This is despite, the fact that the continent consists of women who comprise 51 percent of the population. Thus, with such norms patriarchy restricts women's ability to inherit land (Mapedza, 2024). In addition, accessibility is important, for one to gain access to other common pool resources. For example, access to water for irrigation is sometimes subject to the amount of land owned and the type of crop grown within the vicinity for women. This is because to have one you must have the other. Traditional rules on property rights make it hard for females to gain water supply for irrigation farming as legislation across Africa varies with regards water accessibility and land rights. As water resources are often linked to land ownership (Mapedza, 2024). We must acknowledge how customary land laws affect accessibility for women in the Global South and how this accessibility affects their production and productivity. Many authors acknowledge, that women grow food to feed the family and not for commercial purposes. However, for women to change their status with regards to accessibility they need to receive support from men from their communities, hence, the important connotation of gender equity and not equality. In addition, for women to gain access there must be support from the government to help facilitate this gradual change as some communities may not be ready to accept change about women's access. Women play a critical role in food production and distribution, a fact that is acknowledged by male politicians, who still ignore women, when they are refused access to land ownership in rural areas (Masuku et al., 2023). This is an issue that is important and needs to be addressed seriously by governments who are known to politicize the issue. For example, the need to give women subsidized rates when it comes to land registration or joint titling. Furthermore, showcasing the impacts of such initiatives may motivate others to want to do the same. As the new agricultural era begins it becomes important to have women as part of the conversations as these changes impact them for women.

7.4 Financial Access

The goal of women's financial inclusion is to enhance their ability to access financial services and resources and promote their empowerment and economic development (Elouardighi & Oubejja, 2023). This goal is not only critical and valuable, but is also strategic in encouraging women to have more access to credit and other financial services, financial education and training that will help break some of the social and cultural barriers that harbor them from participating fully in the economy. However, accessibility is easier said than done. This is because financial access is available in different forms and is still subject to social and cultural norms within communities and countries. For example, in some households, men may be the key decision makers when it comes to making decisions regarding funds of the home hence, women only observe or are beneficiaries of these decisions. Women will often follow the decisions made by their partners as social norms within their communities often have them accepting the male view point (Asante et al., 2023). This traditional rules of social behaviour of men and women need to be always carefully assessed. In addition, women on some fronts find it difficult to gain access to finance as they are unable to understand the products and services that are being offered to them and how they can use them. Thus, need does not necessarily translate into access and access does not necessarily translate into use. Finance is the most difficult resource to be accessed by women (Simba et al., 2023). Gender bias within the financial sector, in Africa, continues against women as shown by evidence. Gender plays a central role in determining whether women can or cannot access credit. Furthermore, it limits their ability to draw on business support (Elouardighi & Oubejja, 2023).

7.5 Education Access

It is said that knowledge is power. The importance of women having information is key for them to be able to make key decisions with regard to issues that affect them and in learning how to mitigate situations that affect them. Women have made significant contributions to the agricultural sector and education in Africa. If this continues, this will lead to improved livelihoods and food security with the continent (Yu & Osabohien, 2023). For example, for women to be able to become better producers of certain crops they need training. This is because new farming methods and technologies are on the rise, and the importance of knowing how to produce these crops is important. Discussions on how to plant hybrid varieties, or how to nurture certain livestock are key if farm production levels are to rival those of men. Effective interventions, such as training and education also support women's empowerment (Quisumbing et al., 2023). Increasing women's skills in animal breeding and other forms of knowledge will make them better farmers and increase the same knowledge within their households which in turn will have a direct impact on their ability to access services and make decisions. Authors note that productivity levels in farms by women can be similar to that of men if not higher if only availability of certain infrastructures are available to them: in that regard training opportunities, as well as mechanized tools that can assist them. On the other hand, this can be difficult if women do not have some form of basic education that helps them use the tools they need. In addition, the absence of adequate agricultural extension officers who are trained on gender mainstreaming further makes it difficult for women to learn and exacerbates the issue as women continue to be overlooked (Yee et al., 2024). These challenges make it difficult for women to participate in various social actions that can increase their agency and decrease their vulnerability. Furthermore, it continues the vicious cycle of disempowerment. When women have equal access to production tools as men, their agricultural production levels can increase by 12 to 17 percent and thus reduce global food insecurity. By equipping women with the required tools and knowledge, one empowers them with agency and gives them the power to participate fully in agricultural production (Yee et al., 2024). It becomes very important for agricultural departments to understand the important role that agricultural extension services play in helping to empower farmers, specifically women, looking at cultural contexts and good understanding of the intricacies between resource access, agency, education and achievements (Yee et al., 2024).

7.6 Conclusion

The agrarian revolution is built on three key social concepts within the Social Innovation (SI) framework: people (social), the planet (environment), and production (food). Africa's agricultural revolution has been influenced by previous transformations in Europe and Asia. However, Africa's unique social and cultural norms necessitate a localized approach to achieving meaningful progress in the sector. To drive significant change, it is essential to involve the most critical social group: women. Women are key actors in Africa's food system: they grow, prepare, and serve the food consumed by millions every day. As noted by various scholars, women play a crucial role in revolutionizing Africa's agricultural sector. However, despite their importance, they face numerous constraints that hinder their ability to engage in both subsistence and commercial farming. One of the biggest challenges women face is limited access to land, a fundamental resource for farming. Cultural and gender norms often restrict their ability to own or manage land, limiting their capacity to grow food and maintain nutritious diets for their families. The most impactful change would be for men to actively support women in land ownership specifically by allowing women to co-own land and participate in decision-making regarding land use. More importantly, governments and political leaders must take a firm stance in advocating for women's land ownership and management rights. Land is a critical asset that serves as collateral for securing financial support. Without land ownership, women struggle to access loans that could provide essential inputs such as seeds and fertilizers key resources for improving agricultural productivity. The next generation of farming will be driven by new farming methods and digital tools designed to enhance production. However, women's inability to secure access to these tools hinders their agricultural productivity, thereby creating challenges within the food system. Since, this new farming technology relies heavily on mechanized tools and digital innovations, knowledge and training are essential for farmers. Unfortunately, many women in Africa have only basic education or, in some cases, none at all, depending on their location and cultural customs. This lack of education makes it difficult for them to learn and adopt to the Internet of Things (IoT)-based agricultural tools, as they are more familiar with traditional farming methods. Gender will have to be recognized as a critical factor, if farming in Africa is to succeed. The movement must address the diverse challenges within the industry, particularly those faced by women. Most importantly, this revolution must shift the perception of women in agriculture from mere participants to catalysts of change, driving the sector towards sustainable development and food security.

References

Abdisa, T., Mehare, A., & Wakeyo, M. B. (2024). Analyzing gender gap in agricultural productivity: Evidence from Ethiopia. *Journal of Agriculture and Food Research, [online] 15*, 100960. https://doi.org/10.1016/j.jafr.2023.100960

Abdulai, A. M. (2022). Could eco-feminists or female environmentalists boost the fight against climate change? *African Journal of Climate Change and Resource Sustainability, 1*(1), 107–114.

Achuo, E. D., Asongu, S. A., & Tchamyou, V. S. (2022). *Women empowerment and environmental sustainability in Africa*. Working Paper No. 22/004. The Association for Promoting Women in Research and Development in Africa (ASPROWORDA).

Acosta, M., van Bommel, S., van Wessel, M., Ampaire, E. L., Jassogne, L., & Feindt, P. H. (2019). May discursive translations of gender mainstreaming norms: The case of agricultural and climate change policies in Uganda. In *Women's studies international forum* (Vol. 74, pp. 9–19). Pergamon.

Adefare, T., Adeola, O., Mogaji, E., Nguyen, N. P., & Mogaji, S. A. (2024). Empowering women agriculture entrepreneurs: Banks' role in achieving sustainable development goals. *International Journal of Bank Marketing*.

Adeniyi, L. (2010). Women farmer's and agriculture growth: Challenge and perspective for Africa face the economic crisis.

Adeosun, O. T., & Owolabi, K. E. (2021). Gender inequality: Determinants and outcomes in Nigeria. *Journal of Business and Socio-economic Development, 1*(2), 165–181. https://doi.org/10.1108/JBSED-01-2021-0007

M. O. Jimoh and S. Narsiah, *Unveiling the Power of Women in Irrigation*, https://doi.org/10.1007/978-3-031-94174-0

Adeyanju, D., Mburu, J., Gituro, W., Chumo, C., Mignouna, D., & Mulinganya, N. (2023). Impact of agribusiness empowerment interventions on youth livelihoods: Insight from Africa. *Heliyon, [online] 9*(11), e21291. https://doi.org/10.1016/j.heliyon.2023.e21291

AfDB. (2013). Investing in gender equality for Africa's transformation. https://www.afdb.org/fileadmin/uploads/afdb/Documents/Policy-Documents/2014-2018_-_Bank_Group_Gender_Strategy.pdf

African Capacity Building Foundation (ACBF). (2012). *Africa capacity indicators 2012: Capacity development for agricultural transformation and food security*. ACBF.

Agbo, F. U., Iroh, I. I., & Ihemezie, E. J. (2015). Access to credit by vegetable farmers in Nigeria: A case study of Owerri Agricultural Zone of Imo State, Nigeria. *Asian Journal of Agricultural Research, 9*(4), 155–165.

Agbor, U. I., & Eteng, F. O. (2018). Challenges of rural women in agricultural production and food sufficiency in Cross River State, Nigeria. *Advances in Social Sciences Research Journal, 5*(12).

Agholor, A. I. (2019). Gender gap in Sub-Saharan Africa, reminiscence of rural extension and advisory services: Delineation, challenges and strategies. *South African Journal of Agricultural Extension (SAJAE), 47*(3).

AGRA. (2018). *Africa agriculture status report 2018: Catalyzing government capacity to drive agricultural transformation*. https://agra.org/. Alliance for a Green Revolution in Africa.

Agwata, Jones, F., Gathangu, T. W., & Mulwa, R. M. (2014). Social economic factors hindering the participation of women in managing water resources in Kajiado County, Kenya. *Journal of Humanities and Social Science, 19*(1), 75–85.

Akpa, A. F., Amegnaglo, C. J., & Chabssou, A. F. (2024). Women's engagement in agriculture and income inequality in Sub-Saharan Africa. *Social Sciences and Humanities, 9*, 1–8. https://doi.org/10.1016/j.ssaho.2024.100888

Akpan-Obong, P. (2023). Invisible but ubiquitous: Leveraging ICTs for development in gendered systems of exclusion-Nigeria and Cameroon. In *Handbook of gender and technology* (pp. 56–76). Edward Elgar Publishing.

Akshay, V. R. (2024). *How slar-powered water treatment transforms communities*. [online] Republic of Solar. Available at: https://arka360.com/ros/solar-powered-water-treatment-transforming-communities/

Akter, S., Rutsaert, P., Luis, J., Htwe, N. M., San, S. S., Raharjo, B., & Pustika, A. (2017). Women's empowerment and gender equity in agriculture: A different perspective from Southeast Asia. *Food Policy, 69*, 270–279.

Alkhaled, S., & Karin B. (2018). 'And now I'm free': Women's empowerment and emancipation through entrepreneurship in Saudi Arabia and Sweden. *Entrepreneurship & Regional Development, 30*, 877–900.

Alkire, S., & Sumner, A. (2013). Multidimensional poverty and the post-2015 MDGs. *Development, 56*(1), 46–51.

Alkire, S., Meinzen-Dick, R., Peterman, A., Quisumbing, A., Seymour, G., & Vaz, A. (2012). *The women's empowerment in agriculture index (OPHI Research in Progress 35a)*. Oxford Poverty and Human Development Initiative (OPHI), University of Oxford.

Aluko, B. T., & Amidu. A. (2006). Women and land rights reforms in Nigeria. 5th FIG regional conference, on promoting land administration and good governance. Accra, Ghana.

Ambler, K., Jones, K., & O'Sullivan, M. (2021). Facilitating women's access to an economic empowerment initiative: Evidence from Uganda. *World Development, 138*, 105224. https://doi.org/10.1016/j.worlddev.2020.105224

Amenyah, I. D., & Puplampu, K. P. (2013). *Women in agriculture: An assessment of the current state of affairs in Africa* (pp. 1–45). The African Capacity Building Foundation.

Ampaire, E. L., Acosta, M., Huyer, S., Kigonya, R., Muchunguzi, P., Muna, R., & Jassogne, L. (2020). Gender in climate change, agriculture, and natural resource policies: Insights from East Africa. *Climatic Change, 158*(1), 43–60.

Amusan, L., & Olutola, O. (2017). Contextualizing African women's empowerment in agriculture: Challenges from climate change and mineral extraction perspective's. *Journal of International Women's Studies, 18*(4), 117–130.

Anderson, C. L., Reynolds, T. W., Biscaye, P., Patwardhan, V., & Schmidt, C. (2020). Economic benefits of empowering women in agriculture: Assumptions and evidence. *The Journal of Development Studies, 57*(2), 1–16. https://doi.org/10.1080/00220388.2020.1769071

Asadullah, M. N., & Kambhampati, U. (2021). Feminization of farming, food security and female empowerment. *Global Food Security, 29*, 100532. https://doi.org/10.1016/j.gfs.2021.100532

Asante, B. O., Puskur, R., Garner, E., Mangheni, M. N., Adabah, R., Asante, M. D., Frimpong, B. N., & Prah, S. (2023). Access and control of resources and participation in rice-breeding activities among men and women farmers in Southern Ghana. *Sustainability, 15*(9), 7069. https://doi.org/10.3390/su15097069

Asking, V., & Nilsson, J. (2019). *Risk acceptance in flood affected areas in Nyando, Kenya*. Lund University.

Asongu, S., & Odhiambo, N. (2020). Inequality and the economic participation of women in Sub-Saharan Africa: An empirical investigation. *African Journal of Economic and Management Studies, 11*(2), 193–206.

Ayuk-Etang, E. (2021). Women and eco-spirituality in Achebe's Things Fall Apart (1958): A symbiotic relationship with nature. *Journal of Postcolonial Writing and World Literatures, 2*(2). Retrieved from https://royalliteglobal.com/worldliteratures/article/view/578

Badiane, A., Sané, T., & Thior, M. (2019). *European Scientific Journal, 15*(21), 489–506.

Balabantaray, Dr. S. R. (2023). Women's leadership and sustainable environmental initiatives: A macroscopic investigation from ecofeminism framework. *International Journal of Multidisciplinary Research and Growth Evaluation, 4*(4), 1039–1046.

Barnes, J. (2017). States of maintenance: Power, politics, and Egypt's irrigation infrastructure. *Environment and Planning D: Society and Space, 35*(1), 146–164.

Barrett, C. B., Reardon, T., & Webb, P. (2001). Nonfarm income diversification and household livelihood strategies in rural Afuca: Concepts, dynamics and policy implications. *Food Policy, 26*(4), 315–331.

Benería, L., Berik, G., & Floro, M. (2015). *Gender, development and globalization: Economics as if all people mattered*. Routledge.

Bizoza, A. R. (2019). Land rights and economic resilience of rural women in the G5- sahel countries, West Africa. *African Journal on Land Policy and Geospatial Sciences, 2*(1), 46–59.

Boone, C. (2019). Legal empowerment of the poor through property rights reform: Tensions and trade-offs of land registration and titling in Sub-Saharan Africa. *The Journal of Development Studies, 55*(3), 384–400.

Briggs, R. C. (2015). The influence of aid changes on African election outcomes. *International Interactions, 41*(2), 201–225.

Bryan, E., & Garner, E. (2022). Understanding the pathways to women's empowerment in Northern Ghana and the relationship with small-scale irrigation. *Journal of Agriculture and Human Values, 39*, 905–920.

Budlender, D., & Alma, E. (2011). *Women and land: Securing rights for better lives* (110 pp). International Development Research Centre.

Buehren, N. (2023). *Gender and agriculture in Sub-Saharan Africa: Review of constraints and effective interventions*. Gender Innovation Lab, World Bank.

Buvinic, M., & o'Donnell, M. (2016). *Revisiting what works: Women, economic empowerment and smart design*. Center for Global Development.

Buvinic, M., O'Donnell, M., & Bourgault, S. (2020). *Women's economic empowerment in West Africa: Towards a practical research agenda* (557 pp). Center for Global Development Working Paper.

Caliyurt, K. (2016). *Women and sustainability in business: A global perspective*. Taylor & Francis.

Calkin, S. (2015). Feminism, interrupted? Gender and development in the era of 'Smart Economics'. *Progress in Development Studies, 15*(4), 295–307.

Cameron, E., Trend-Cunningham, F., Ray, J. M., & Jacquin, K. M. (2022). *Indicators of climate change and violence against women predict estimated prevalence of modern slavery: An ecofeminist perspective*. Research Square.

CGIAR (Consultative Group for International Agricultural Research) research program on Water, Land and Eco-systems (WLE) (2018). *Gender equitable pathways to achieving sustainable agricultural intensification*. International Water Management Institute.

Chauhan, R., & Chauhan, J. (2015). Women: Seeds of change in agriculture. *Indian Research Journal of Extension Education, 15*(3), 72–79.

Chege, C. G. K., Onyango, K., Muange, E. N., & Jäger, M. (2021). *How women's empowerment affects farm production and dietary quality in East Africa* (24 p). CIAT Publication No. 520. International Center for Tropical Agriculture (CIAT).

Chigbu, U., Paradza, G., & Dachaga, W. (2019). Differentiations in women's land tenure experiences: Implications for women's land access and tenure security in Sub-Saharan Africa. *Land, 8*(2), 22.

Chigusiwa, L., Kembo, G., & Kairiza, T. (2023). Drought and social conflict in rural Zimbabwe: Does the burden fall on women and girls? *Review of Development Economics, 27*(1), 178–197.

Chitando, E., Conradie, E. M., & Kilonzo, S. M. (2022). *African perspectives on religion and climate change* (p. 254). Taylor & Francis.

Christophe B., & Devereux, S. (2023). *Resilience and food security in a food systems context*. Springer Nature.

Cornwall, A., & Rivas, A. M. (2015). From 'gender equality and 'women's empowerment to global justice: Reclaiming a transformative agenda for gender and development. *Third World Quarterly, 36*(2), 396–415.

Dancer, H. (2017). An equal right to inherit? Women's land rights, customary law and constitutional reform in Tanzania. *Social & Legal Studies, 26*(3), 291–310.

Das, S. (2016). From eco-sensibility to ecofeminism. In *Re-thinking environment: Literature, ethics and praxis* (pp. 49–59). Authors Press.

De Luca, C. (2018). Global Crisis, Development and the Emergence of Women Voices = Crisis global, desarrollo y la aparición de las voces de las mujeres. *FEMERIS: Revista Multidisciplinar de Estudios de Género, 3*(1), 35–48. https://doi.org/10.20318/femeris.2018.4071

Devi, P., & Kumar, S. (2023). Two sustainable pillars of development: Women and nature from an ecofeminist perspective. *The Criterion: An International Journal in English, 14*(5), 1–9.

Devkota, R., Pant, L. P., Hambly Odame, H., Rai Paudyal, B., & Bronson, K. (2022). Rethinking gender mainstreaming in agricultural innovation policy in Nepal: A critical gender analysis. *Agriculture and Human Values*. https://doi.org/10.1007/s10460-022-10326-1

Diiro, G. M., Seymour, G., Kassie, M., Muricho, G., & Muriithi, B. W. (2018). Women's empowerment in agriculture and agricultural productivity: Evidence from rural maize farmer households in western Kenya. *PLOS ONE, 13*(5).

Djurfeldt, A. A., Dzanku, F. M., & Isinika, A. C. (2018a). Perspectives on agriculture, diversification, and gender in Rural Africa: Theoretical and methodological issues. In *Agriculture, diversification, and gender in rural Africa: Longitudinal perspectives from six countries* (Vol. 1). Oxford University Press.

Djurfeldt, A. A., Hillbom, E., Mulwafu, W. O., Mvula, P., & Djurfeldt, G. (2018b). 'The family farms together, the decisions, however are made by the man'—Matrilineal land tenure systems, welfare and decision making in rural Malawi. *Land Use Policy, 70*, 601–610.

Dolan, C. (2001). The good wife's struggle over resources in the Kenyan horticultural sector. *The Journal of Development Studies, 37*(3), 39–70.

Dos Santos, T. (1998) 'The structure of dependence'. In Frank, A. G., Sunkel, O., & Gordon, R. F. (eds.) *Dependency and development in Latin America*. Routledge.

Doss, C. (2015). *Women and agricultural productivity: What does the evidence tell us?* Economic Growth Center, Yale University.

Doss, C., Meinzen-Dick, R., Quisumbing, A., & Theis, S. (2018). Women in agriculture: Four myths. *Global Food Security, 16*, 69–74.

Doss, C., Summerfield, G., & Tsikata, D. (2014). "Land, Gender, and Food Security." *Feminist Economics, 20*, 1–23.

Doss, C. R. (2001). Designing agricultural technology for African women farmers: Lessons from 25 years of experience. *World Development, 29*(12), 2075–2092.

Doss, C. R. (2018). Women and agricultural productivity: Reframing the Issues. *Development Policy Review, 36*(1), 35–50.

Doss, C. R., & Agnes R. Q. (2020). "Understanding rural household behavior: Beyond Boserup and Becker". *Agricultural Economics, 51*(1), 47–58.

Duflo, E. (2012). Women empowerment and economic development. *Journal of Economic Literature, 50*(4), 1051–1079.

Efobi, U. R., Tanaken, B. V., & Asongu, S. A. (2018). Female economic participation with information and communication technology advancement: Evidence from Sub-Saharan Africa. *South African Journal of Economics, 86*(2), 231–246. https://doi.org/10.1111/saje.12194

Egbai, M. J. (2018). Ecofeminism and political communication: A philosophical assessment of Wangari Maathai's contribution.

Egwurube, J. (2016). Challenges facing women empowerment in contemporary Nigeria. *Miroirsevue des civilisations anglophone, ibérique et ibéro-américaine, 2*(4), 134–168.

Elouardighi, I., & Oubejja, K. (2023). Can digital financial inclusion promote women's labor force participation? Microlevel evidence from Africa. *International Journal of Financial Studies, 11*, 87. https://doi.org/%2010.3390/ijfs11030087

Emmanuel, D., Owusu-Sekyere, E., Owusu, V., & Jordaan, H. (2016). Impact of agricultural extension service on adoption of chemical fertilizer: Implications for rice productivity and development in Ghana. *NJAS - Wageningen Journal of Life Sciences, 79*, 41–49.

Enaifoghe, A., Brown, C., Maramura, T. C., & Ajede, S. A. (2021). Bridging the inequality gap in Africa for sustainable development: Assessing the role of women. *International Journal of Development and Sustainability, 10*(7), 264–279.

Fabiyi, E. F., & Akande, K. E. (2015). Economic empowerment for rural women in Nigeria: Poverty alleviation through agriculture. *Journal of Agricultural Science, 7*(9), 236–241.

FAO. (2013). *Climate-smart agriculture sourcebook.* Rome.

FAO. (2019). *Country, gender, assessment of agriculture and rural sector in Nepal*. Food and Agricultural Organisation.

Food and Agriculture Organization of the United Nations (FAO). (2023). The status of women in agrifood systems—Overview. Rome. https://doi.org/10.4060/cc5060en

Farhall, K., & Rickards, L. (2021). The 'Gender Agenda' in agriculture for development and its (lack of) alignment with feminist scholarship. *Frontiers in Sustainable Food Systems, 5*.

Fernandez, A., Della Giusta, M., & Kambhampati, U. S. (2015) The intrinsic value of agency: The case of Indonesia. *World Development, 70*(C), 92–107.

Fischer, E., & Qaim, M. (2012). Gender, agricultural commercialization, and collective action in Kenya. *Food Security, 4*(3), 441–453. https://doi.org/10.1007/s12571-012-0199-7

Fitawek, W., Hendriks, S., Reys, A., & Fossi, F. (2020). The effect of large-scale agricultural investments on household food security in Madagascar. *Food Security, 12*(6), 1349–1365. https://doi.org/10.1007/s12571-020-01055-6

Fonjong, L. N., & Gyapong, A. Y. (2021). Plantations, women, and food security in Africa: Interrogating the investment pathway towards zero hunger in Cameroon and Ghana. *World Development, 138*, 105293. https://doi.org/10.1016/j.worlddev.2020.105293

Gaard, G. C., & Patrick D. M. (1998). *Ecofeminist literary criticism: Theory, interpretation, pedagogy* (vol. 13). University of Illinois Press.

Gaard, G. ed. (1993). *Ecofeminism: Women, animals, nature.* Temple University Press.

Gaebl, G. (2018). Acta cogitata: An undergraduate journal in philosophy on the liberation of all women: Socialist feminism and materialist ecofeminism recommended citation. *Undergraduate Journal in Philosophy, 5*(2), 2–8.

Galiè, A., Mulema, A., Mora Benard, M. A., Onzere, S. N., & Colverson, K. E. (2015). Exploring gender perceptions of resource ownership and their implications for food security among rural livestock owners in Tanzania, Ethiopia, and Nicaragua. *Agriculture & Food Security, 4*(1), 1–14. https://doi.org/10.1186/s40066-015-0021-9

Gashu, D., Demment, M. W., & Stoecker, B. J. (2019). Challenges and opportunities to the African agriculture and food systems. *African Journal of Food Agriculture Nutrition and Development, 19*(1), 14190–14217. https://doi.org/10.18697/ajfand.84.BLFB2000

Gbigbi, T. M. (2019). Why worry about informal money lenders patronage? A research agenda: Evidence from delta state, Nigeria. *Black Sea Journal of Agriculture, 2*(2), 71–78.

Gebreyohannes, N. M., & David, A. D. (2022). Women and nature: An ecofeminist reading of chimamanda ngozi adichie's purple hibiscus. *Literature, 2*(3), 179–188. https://doi.org/10.3390/literature2030015. MDPI.

Gella, A. A., & Tadele, G. (2014). *Gender and farming in Ethiopia: An exploration of discourses and implications for policy and research* (pp. 1–16). The Future Agricultures Consortium.

Ghasemi, M., Badsar, M., Falahati, L., & Karamidehkordi, E. (2021). The mediation effect of rural women empowerment between social factors and environment conservation (combination of empowerment and eco-feminist theories). *Environment, Development and Sustainability, 23*, 13755–13777. https://doi.org/10.1007/s10668-021-01237-y

Ghosh, T., & Ramanayake, S. S. (2018). *Women empowerment and good times: Which one leads to the other?* Indra Gandhi Institute of Development Research.

Giacomini, T., Turner, T., Isla, A., & Brownhill, L. (2018). Ecofeminism against capitalism and for the commons. *Capitalism Nature Socialism, [online] 29*(1), 1–6. https://doi.org/10.1080/10455752.2018.1429221

Glauber, J., & Laborde, D. (2023). *Repurposing food and agricultural policies to deliver affordable healthy diets, sustainably and inclusively: What is at stake? Background paper for The State of Food Security and Nutrition in the World 2022.* Food & Agriculture Org.

Glazebrook, T., Noll, S., & Opoku, E. (2020). Gender matters: Climate change, gender bias, and women's farming in the global south and north. *Agriculture, 10*(7), 267. https://doi.org/10.3390/agriculture10070267

GOK. (2021). *Ministry of agriculture, livestock, fisheries and co-operatives: Food, our health, wealth, and security agricultural policy-2021.* Government of Kenya.

Gordon, J. (2016). Globalization from below: An ecofeminist analysis of the importance of food security. *University of Guelph's Undergraduate Feminist Journal, 2*, 52–62.

Gottlieb, J., Grossman, G., & Robinson, A. L. (2016). Do men and women have different policy preferences in Africa? Determinants and implications of gender gaps in policy prioritization. *British Journal of Political Science, 48*(3), 611–636. https://doi.org/10.1017/s0007123416000053

Government of Kenya (GoK). (2010). *The constitution of Kenya 2010*. Government Printer.

Gowda, N. D. (2018). An analysis on the role of women in agriculture and rural development in the Indian context. *International Journal of Legal Research and Studies, 3*(1), 127–133.

Grantham, K. E., Dowie, G., & Haan, A. de (2021). *Women's economic empowerment: Insights from Africa and South Asia*. Routledge; International Development Research Centre.

Gumucio, T., Hansen, J., Huyer, S., & van Huysen, T. (2019). Gender-responsive rural climate services: A review of the literature. *Climate and Development, 12*(3), 241–254.

Habtezion, S. (2016). Overview of linkages between gender and climate change. Training module 1. United Nations Development Programme. UNDP.

Hailemariam, A., Kalsi, J., & Mavisakalyan, A. (2024). Gender gaps in the adoption of climate-smart agricultural practices: Evidence from Sub-Saharan Africa. *Journal of Agricultural Economics, 75*(2), 764–793. https://doi.org/10.1111/1477-9552.12583

Halim, S. (2015). Women's Ownership Rights and Access to Land in Bangladesh. In Kelkar, G., & Krishnaraj, M. (eds.) *Women, land and power in Asia*. Routledge India.

Harrison, E., & Mdee, A. (2017). Successful small-scale irrigation or environmental destruction? The political ecology of competing claims on water in the Uluguru Mountains,Tanzania. *Journal of Political Ecology, 24*(1), 406–424.

Hart, T., & Aliber, M. (2015). *Inequalities in agricultural support for women in South Africa. Hsrc.ac.za*. Human Science Research Council. http://hdl.handle.net/20.500.11910/3235

Hausermann, H., Ferring, D., Atosona, B., Mentz, G., Amankwah, R., Chang, A., Hartfield, K., Effah, E., Asuamah, G. Y., Mansell, C., & Sastri, N. (2018). Land-grabbing, land-use transformation and social differentiation: Deconstructing 'small-scale' in Ghana's recent gold rush. *World Development, 108*, 103–114.

Hebnick, P., Bosma, L., & Gert Jan, V. (2019). Petrol pumps and the making of modernity along the shores of Lake Victoria, Kenya. *Water-alternatives.org, 12*(1), 13–29.

Hellin, J., & Fisher, E. (2018). Building pathways out of poverty through climate smart agriculture and effective targeting. *Development in Practice, 28*(7), 974–979. https://doi.org/10.1080/09614524.2018.1492516

Hickey, S., Lavers, T., Seekings, J., & Nino Zarazua, M. (2018). *The negotiated politics of social protection in Sub-Saharan Africa. WIDER Working Paper Series*. World Institute for Development Economic Research (UNU-WIDER).

Hlahla, S. (2022). Gender perspectives of the water, energy, land, and food security nexus in Sub-Saharan Africa. *Frontiers in Sustainable Food Systems, 6*, 719913.

Hornset, N., & de Soysa, I. (2021). Does empowering women in politics boost human development? An empirical analysis, 1960–2018. *Journal of Human Development and Capabilities*, 1–28.

Hosseinnezhad, F. (2017). Women and the environment: Ecofeministic approach to environmental attitudes and Behavior in Iran. *European Journal of Sustainable Development Research, 1*(1), 1–7

Howland, F., Le Coq, J.-F., & Acosta, M. (2019). *Gender integration in agriculture, food security and climate change policy: A framework proposal* (pp. 1–35). CGIAR.

Huyer, S. (2016). Closing the gender gap in agriculture. *Gender, Technology and Development, 20*(2), 105–116. https://doi.org/10.1177/0971852416643872

Ingutia, R., & Sumelius, J. (2021). Do farmer groups improve the situation of women in agriculture in rural Kenya? *International Food and Agribusiness Management Review, 25*(1), 1–22.

IWMI. IWMI Gender and Inclusion Strategy 2020–2023: New Landscapes of Water Equality and Inclusion; International Water Management Institute (IWMI): Sri Lanka, Colombo, 2020; Available online https://www.iwmi.cgiar.org/about/iwmi-genderand-inclusion-strategy-2020-2023/ (accessed on 20 August 2022).

Jacobs, S. (2014). Gender, land and sexuality: Exploring connections. *International Journal of Politics, Culture and Society, 27*(2), 173–190.

Jacob, A. S. (2017). The politics of land and the politics of Gender: Sarah Joseph's gift in Green. *The Criterion: An international Journal in English, 8*(2), 75–83.

Jaiyeola, E. F., & Adeyeye, M. M. (2021). Obstacles along the path of women enterprises in Africa: A case study of Ogotun women in Ekiti state, Nigeria. *Heliyon, 7*(7), e07593. https://doi.org/10.1016/j.heliyon.2021.e07593

Jaquette, J. S. (2017). Women/Gender and Development: The growing gap between theory and practice. *Studies in Comparative International Development, [online] 52*(2), 242–260. https://doi.org/10.1007/s12116-017-9248-8

Jayne, T. S., Muyanga, M., Wineman, A., Ghebru, H., Stevens, C., Stickler, M., Chapoto, A., Anseeuw, W., Westhuizen, D., & Nyange, D. (2019). Are medium-scale farms driving agricultural transformation in Sub-Saharan Africa? *Agricultural Economics, 50*(S1), 75–95. https://doi.org/10.1111/agec.12535

Jellason, N. P., Robinson, E. J. Z., & Ogbaga, C. C. (2021). Agriculture 4.0: Is Sub-Saharan Africa ready? *Applied Sciences, 11*(12), 5750. https://doi.org/10.3390/app11125750

Jimbira, S. S., & Hathie, I. (2020). *The future of agriculture in Sub-Saharan Africa policy brief No:2*. Southern Voice, p. 18.

Johansson, K. (2016). *How integrated are women and gender in integrated water resource management?* Thesis.

Jones, R. (2017). Sustainability: An ecofeminist perspective. *The Student Researcher, 4*(1), 61–67.

Julien, J. C., Bravo-Ureta, B. E., & Rada, N. E. (2023). Gender and agricultural productivity: Econometric evidence from Malawi, Tanzania, and Uganda. *World Development, 171*, 106365.

Kaaria, S., Osorio, M., Wagner, S., & Gallina, A. (2016). Rural women's participation in producer organizations: An analysis of the barriers that women face and strategies to foster equitable and effective participation. *Journal of Gender, Agriculture and Food Security, 1*(2), 148–167.

Kabeer, N. (2021). Gender equality, inclusive growth, and labour markets. In *Women's economic empowerment* (pp. 13–48). Routledge.

Kadzamira, M., Chege, F., Suntharalingam, C., Bundi, M., Likoko, L., Magero, D., Romney, D., Kansiime, M., & Mulema, J. (2024). African women and young people as agriculture service providers—business models, benefits, gaps and opportunities. *CABI Agriculture and Bioscience, 5*(1), 1–14.

Kahan, D., Bymolt, R., & Zaal, F. (2018). Thinking outside the plot: Insights on small-scale mechanisation from case studies in East Africa. *The Journal of Development Studies, 54*(11), 1939–1954.

Kamande, M. W., & Bahati, E. M. (2019). Effects of women land rights on agricultural outcomes in Rwanda. *Journal of Land Policy and Geospatial Sciences, 2*(1), 60–70.

Kansiime, M. K., Alawy, A., Allen, C., Subharwal, M., Jadhav, A., & Parr, M. (2019). Effectiveness of mobile agri-advisory service extension model: Evidence from Direct2Farm program in India. *World Development Perspectives, 13*, 25–33. https://doi.org/10.1016/j.wdp.2019.02.007

Kelkar, G., & Govindnathan, P. (2024). Social norms and attitudes towards women's entitlement to land. *India Social Development Report 2023*, 59.

Kernecker, M., Vogl, C. R., & Aguilar Meléndez, A. (2017). Women's local knowledge of water resources and adaptation to landscape change in the mountains of Veracruz, Mexico. *Ecology and Society, 22*(4).

Khanduja, G. (2017). Prakriti and Shakti: An ecofeminist perspective. *Jindal Journal of Public Policy, 3*(1), 105–114.

Khosrobeigi Bozchelouie, R. (2018). Gender identities and farm survival: Women's activities in the agricultural sector in remote rural areas—Corvinus Research Archive. *Corvinus Journal of Sociology and Social Policy, 9*(2), 77–98.

Kirui, O. K. (2019). *The complementarity of education and productive inputs among smallholder farmers in Africa*. University of Bonn, Center for Development Research (ZEF).

Kisang, E., & Yator, F. (2017). Factors influencing rice production in Kenya's irrigation schemes: The case of Mwea irrigation scheme. *Mara Research Journal of Business & Management, 1*(1), 107–114.

Kisumu County Urban Sustainability Review. (2017). *Kisumu county urban sustainability review; technical working group; Symbio city approach to sustainable urban development*. Kisumu County.

Knutsen, T., & Kotsadam, A. (2020). The political economy of aid allocation: Aid and incumbency at the local level in Sub Saharan Africa. *World Development, 127*, 104729.

Komakech, C. H. (2013). Emergence and evolution of endogenous water institutions in an African river basin. Local water governance and state intervention in the Pangani riverbasin, Tanzania. Dissertation submitted in fulfillment of the requirements of the Board for Doctorates of Delft University of Technology and the Academic Board of the UNESCOIHE Institute for Water Education for the Degree of Doctor. CRC Press/Balkema.

Kote, P., Yallappa, M., Jabeen, A., Srinatha, T. N., Prabhavathi, S.J., Ramasamy, M., Dhanalakshmi, K., Chitra, K., & Malathi, G. (2024). A scoping review on youth participation in agriculture: Sustainable development, food security, and economic growth. *Journal of Scientific Research and Reports, 30*(5), 947–958. https://doi.org/10.9734/jsrr/2024/v30i52012

Kristjanson, P., Bryan, E., Bernier, Q., Twyman, J., Meinzen-Dick, R., Kieran, C., Ringler, C., Jost, C., & Doss, C. (2017). Addressing gender in agricultural research for development in the face of a changing climate: Where are we and where should we be going? *International Journal of Agricultural Sustainability, 15*(5), 482–500. https://doi.org/10.1080/14735903.2017.1336411

Kwauk, C., & Braga, A. (2017). *BROOKE SHEARER SERIES Three platforms for girls' education in climate strategies* (pp. 1–52). The Brookings Institution.

Laborde, D., Martin, W., Swinnen, J., & Vos, R. (2020). COVID-19 risks to global food security. *Science, 369*(6503), 500–502.

Lambrecht, I., Schuster, M., Asare Samwini, S., & Pelleriaux, L. (2018). Changing gender roles in agriculture? Evidence from 20 years of data in Ghana. *Agricultural Economics, 49*(6), 691–710.

Lau, J. D., Kleiber, D., Lawless, S., & Cohen, P. J. (2021). Gender equality in climate policy and practice hindered by assumptions. *Nature climate change, 11*(3), 186–192.

Lawry, S., Samii, C., Hall, R., Leopold, A., Hornby, D., & Mtero, F. (2014). The impact of land property rights interventions on investment and agricultural productivity in developing countries: A systematic review. *Journal of Development Effectiveness, 9*(1), 61–81. https://doi.org/10.4073/csr.2014.1

Lawry, S., Samii, C., Hall, R., Leopold, A., Hornby, D., & Mtero, F. (2017). The impact of land property rights interventions on investment and agricultural productivity in developing countries: A systematic review. *Journal of Development Effectiveness, 9*(1), 61–81. https://doi.org/10.1080/19439342.2016.1160947

Le, N. P. (2018). Gender roles in response to water scarcity for coffee production: A case study of an ethnic group in the central highlands of Vietnam. *Journal of Social Sciences, 13*(4), 897–898.

Leach, M., Nisbett, N., Cabral, L., Harris, J., Hossain, N., & Thompson, J. (2020). Food politics and development. *World Development, 134*, 105–024.

Lebel, L., Lebel, P., Manorom, K., & Zhou, Y. (2019). Gender in development discourses of civil society organizations and Mekong hydropower dams. *Water Alternatives, 12*(1), 192–220.

Lecoutere, E., Kosec, K., Quisumbing, A., Elias, M., Bryan, E., & Puskur, R. (2023). *Equality and empowerment by gender and intersecting social differentiation in agri-food systems: Setting the stage*. FAO.

Lee, A. (2018). An ecofeminist perspective on new food technologies. *Canadian Food Studies/La Revue canadienne des études sur l'alimentation, 5*(1), 63–89.

Lefore, N., Weight, E., & Mukhamedova, N. (2017a). *Improving gender equity in irrigation: Application of a tool to promote learning and performance in Malawi and Uzbekistan. Cgiar.org*. International Water Management Institute (IWMI). https://doi.org/2522-7106%20(Print)

Lefore, N., Giordano, M., Ringler, C., & Barron, J. (2019b). Sustainable and equitable growth in farmer-led irrigation in Sub-Saharan Africa: What will it take? *Water Alternatives, 121*, 156–158.

Lekwauwa O., & Ijioma, J. C. (2012). Mainstreaming women and youths in climate change mitigation and adaptation in the less developed countries. Proceedings, 17 th Annual National Conference. Agriculture Extinction Society of Nigeria

Lorentzen, L. A., & Eaton, H. (2002). Ecofeminism: An overview. In *The forum on religion and ecology at yale* (pp. 1–2).

Lv, C., & Wang, Z. (2018). Carolyn merchant's view of nature from the perspective of ecofeminism. *Advances in Social Sciences, Education and Humanities Research, 176*, 1–3.

Macgregor, S. (2017). *Routledge handbook of gender and environment*. Taylor & Francis.

Macharia, A. M., Carpio, C. E., & Kirimi, L. (2018). Heterogeneity in irrigation technology impacts: Implications for adoption. 2018 Agricultural and Applied Economics Association Annual Meeting. Texas Tech University.

Magill, G., & Benedict, J. (2019). *Cascading challenges in the global water crisis*. Cambridge Scholars Publishing.

Mahour, K. (2016). Role of women in environment conservation. *Advances in BioScience, 7*(1), 17–26.

Maïga, E., Christiaensen, L., & Palacios-Lopez, A. (2017). *Are the youth exiting agriculture en masse?* World Bank.

Majiwa, E., & Mugodo, C. (2018). Technical efficiency and technology gap ratios among rice farmers in Kenya. In *Conference of agricultural economists international conference of agricultural economists*. Jomo Kenyatta University of Agriculture and Technology.

Makhado, R., & Pelizzo, R. (2016). Looking at women rights to land in Africa: A feminist perspective? African politics and policy. *Journal of African Politics and Policy, 2*(1), 32–40.

Makini, F., Kamau, G., Mose, L., Ongala, J., Salasya, B., Mulinge, W., & Makelo, M. (2017). Status, challenges, and prospects of agricultural mechanisation in Kenya: The case of rice and banana value chains. *FARA Research Report, 1*(2), 1–24.

Malapit, H. J. L., & Quisumbing, A. R. (2015). 'What dimensions of women's empowerment in agriculture matter for nutrition in Ghana?' *Food Policy, 52*, 54–63. https://doi.org/10.1016/j.foodpol.2015.02.003

Mapedza, E. (2024). Gender norms and social transformation of agriculture in Sub-Saharan Africa. In *Circular and transformative economy* (pp. 159–174). CRC Press.

Mapedza, E., Tagutanazvo, E. M., van Koppen, B., & Manyamba, C. (2017) Agricultural water management in matrilineal societies in Malawi: Landownership and implications for collective action. In Suhardiman, D., Nicol, A., & Mapedza, E. (eds.) *Water governance and collective action—Multi scale challenges* (pp. 82–95). Routledge, Taylor and Francis Group.

Marchand, M. H. (2009). The future of gender and development after 9/11: Insights from postcolonial feminism and transnationalism. *Third World Quarterly, 30*(5), 921–935. https://doi.org/10.1080/01436590902959149

Masaki, T. (2018). The political economy of aid allocation in Africa: Evidence from Zambia. *African Studies Review, 61*(1), 55–82.

Mashapure, R., Nyagadza, B., Chikazhe, L., Mazuruse, G., & Hove, P. (2023). Women entrepreneurship development and sustainable rural livelihoods in Zimbabwe. *Arab Gulf Journal of Scientific Research*. https://doi.org/10.1108/agjsr-07-2022-0112

Masuku, M. M., Mthembu, Z., & Mlambo, V. H. (2023). Gendered effects of land access and ownership on food security in rural settings in South

Africa. *Frontiers in Sustainable Food Systems, 7*. https://doi.org/10.3389/fsufs.2023.1158946

Mateo, S. -J., Zadeh, S. M., & Turral, H. (2018). *More people, more food, worse water? A global review of water pollution from agriculture*. Food and Agriculture Organisation (FAO).

Matheson, M., & Heinze, A. (2019). Reevaluating African women's inheritance rights in indigenous customary law and statutory national law. BYU ScholarsArchive Citation. *BYU-Scholars Archive, 33*, 1–15.

Mathis, A. (2024). Empowering Ugandan women smallholder farmers: Documenting her mighty hands agriculture extension nonprofit.

Mbatha, S., & Pitt, B. (2021). *Black lion. Alive in the wilderness*. Jonathan Balls Publishers.

Mdee, A., Ofori, A., Chasukwa, M., & Manda, S. (2020). Neither sustainable nor inclusive: A political economy of agricultural policy and livelihoods in Malawi, Tanzania and Zambia. *The Journal of Peasant Studies*, 1–24. https://doi.org/10.1080/03066150.2019.1708724

Meinzen-Dick, R., Johnson, N., Quisumbing, A. R., Njuki, J., Behrman, J. A., Rubin, D., Peterman, A., & Waithanji, E. (2014). Gender in agriculture. In *Gender in agriculture: Closing the gender gap*. Springer.

Meinzen-Dick, R., Kovarik, C., & Quisumbing, A. R. (2014). *Gender and sustainability: Annual review of environment and resources* (pp. 29–55). International Food Policy Research Institute. https://doi.org/10.1146/annurev-environ-101813-013240

Meinzen-Dick, R., Quisumbing, A., Doss, C., & Theis, S. (2019). Women's land rights as a pathway to poverty reduction: Framework and review of available evidence. *Agricultural Systems, 172*, 72–82. https://doi.org/10.1016/j.agsy.2017.10.009

Menon, N. (2018). *Gender and development*. Indira Gandhi National Open University.

Mills, G. (2018). Why is Africa poor? In *Culture and civilization* (pp. 52–62). Routledge.

Mishra, N. (2018). *Gender and environment*. Indra Gandhi National Open University.

Mitra, A., & Rao, N. (2019). Gender, water, and nutrition in India: An intersectional perspective - UEA Digital Repository. *Journal of Water Alternatives, 12*(1), 169–191.

Mohapatra, M. B. (2019). Raising the voice against the patriarchal monopoly: Reflecting on the evolution of ecofeminism in the post colonial era. *International Journal of English Language, Literature and Humanities, 7*(1), 1–18.

Mudege, N. N., Mdege, N., Abidin, P. E., & Bhatasara, S. (2017). The role of gender norms in access to agricultural training in Chikwawa and Phalombe, Malawi. *Gender, Place & Culture, 24*(12), 1689–1710.

Muema, F., Home, P., & Raude, J. (2018). Application of benchmarking and principal component analysis in measuring performance of public irrigation schemes in Kenya. *Journal of Agriculture, 8*(10), 1–20.

Mugambi, G., Kihara, P., & Senaji, T. (2017). Analysis of strategic options and performance of irrigation schemes in Kenya: Bura irrigation scheme. *International Journal of Contemporary Aspects in Strategic Management (IJCASM), 1*(2), 27–38.

Mukasa, A. N., Woldemichael, A. D., Salami, A. O., & Simpasa, A. M. (2017). Africa's agricultural transformation: Identifying priority areas and overcoming challenges. *Africa Economic Brief, 8*(3), 1–16.

Mukherjee, R. (2013). Eco-feminism: Role of women in environmental governance and management. *Galgotias Journal of Legal Studies, 1*(2), 1–7.

Mukoni, M., Mudaly, R., & Moletsane, R. (2018). Gender, power and women's participation in community environmental education. *Southern African Journal of Environmental Education, 34*, 15–29.

Muralidharan, A., Fehringer, J., Pappa, S., Rottach, E., Das, M., & Mandal, M. (2014). *Transforming gender norms, roles, and power dynamics for better health: Evidence from a systematic review of gender-integrated health programs in low- and middle-income countries.*

Muraoka, R., Jin, S., & Jayne, T. S. (2018). Land access, land rental and food security: Evidence from Kenya. *Land Use Policy, 70*, 611–622.

Muriithi, B. W. (2018). *Does gender matter in the adoption of sustainable agricultural technologies? A case of push-pull technology in Kenya.* International Centre of Insect Physiology and Ecology (ICIPE).

Mushtaq, R., Dastane, O., Rafiq, M., & Basar, B. D. (2023). "Women financial inclusion research: A bibliometric and network analysis". *International Journal of Social Economics*. https://doi.org/10.1108/IJSE-06-2022-0438

Naguib, R. (2022). Motivations and barriers to female entrepreneurship: Insights from Morocco. *Journal of African Business*, 1–28. https://doi.org/10.1080/15228916.2022.2053400

Nakawuka, P., Langan, S., Schmitter, P., & Barron, J. (2017). A review of trends, constraints and opportunities of smallholder irrigation in East Africa. *Global Food Security, 17*, 196–212.

Namonje-Kapembwa, T., & Chapoto, A. (2016). *Improved agricultural technology adoption in Zambia: Are women farmers being left behind*? Indaba Agricultural Policy Research Institute.

Naples, N. A. (Ed.). (2020). *Companion to women's and gender studies*. Wiley.

National Irrigation Board (n.d.). Ahero Irrigation Scheme. [online] National Irrigation Board of Kenya. Available at: https://nib.or.ke/. Accessed 2 Oct 2018.

Ndeda, M. A. J. (2019). Population movement, settlement and the construction of society to the east of Lake Victoria in precolonial times: The western Kenyan case. *The East African Review,* (52), 83–108.

Nelson, S., & Huyer, S. (2016). A gender-responsive approach to climate smart agriculture: Evidence and guidance for practitioners. In *Climate-smart agriculture practice brief*. CGIAR Research Program on Climate Change, Agriculture and Food Security (CCAFS). https://hdl.handle.net/10568/73049

Newell, P., Taylor, O., & Touni, C. (2017). Governing food and agriculture in a warming world. *Global Environmental Politics, 18*(2), 53–71. https://doi.org/10.1162/glep_a_00456

Ng'uono, M., Maritim, K. K., Okello, G., Onganya, D., & Kuto, L. (2014). Social cultural factors that constraint gender mainstreaming in agriculture extension. *Journal of Social Sciences, 2*, 1–19.

Ng'weno, A., Oldja, L., Hassan, M., Kapoor, P. (2018). *Demand-side review of financial inclusion for women in entrepreneurship and smallholder agriculture.* International Development Research Centre.

Ngomane, T. S., & Sebola, M. P. (2016). Agricultural markets as nodal points for economic activity: Are agricultural markets gender inclusive? In *SAAPAM Limpopo Chapter 5th Annual Conference Proceedings 2016.*

Nhemachena, C., Nhamo, L., Matchaya, G., Nhemachena, C. R., Muchara, B., Karuaihe, S. T., & Mpandeli, S. (2020). Climate change impacts on water and agriculture sectors in Southern Africa: Threats and opportunities for sustainable development. *Water, 12*(10), 2673.

Niaghi, O. (2019). The role of women and their empowerment in environmentally sustainable development. *Journal of Environmental Science Studies, 4*(2), 1328–1339.

Nicol, A., Langan, S., Victor, M., & Gonsalves, J. (2015). *Water-smart agriculture in East Africa.* International Water Management Institute (IWMI). CGIAR Research Program on Water, Land and Ecosystems (WLE); Global Water Initiative East Africa (GWI EA). https://doi.org/10.5337/2015.203

Njuki, J., Melesse, M., Sinha, C., Seward, R., Renaud, M., Sutton, S., Nijhawan, T., Clancy, K., Thioune, R., & Charron, D. (2022). Meeting the challenge of gender inequality through gender transformative research: lessons from research in Africa, Asia, and Latin America. *Canadian Journal of Development Studies/Revue canadienne d'études du développement, 44*(2), 1–23.

Njuki, J., Melesse, M., Sinha, C., Seward, R., Renaud, M., Sutton, S., Nijhawan, T., Clancy, K., Thioune, R., & Charron, D. (2023). Meeting the challenge of gender inequality through gender transformative research: lessons from research in Africa, Asia, and Latin America. *Canadian Journal of Development Studies/Revue canadienne d'études du développement, 44*(2), 206–228.

Njuki, J., Parkins, J. R., & Kaler, A. (2016). *Transforming gender and food security in the global south.* Routledge.

Nkonya, L. (2006). Drinking from own cistern: Customary institutions and their impacts on rural water management in Tanzania. A dissertation submitted in partial fulfillment of the requirements for the degree Doctor of Philosophy Department of Sociology, Anthropology and Social Work College of Arts and Sciences Kansas State University Manhattan Kansas.

Nkonya, L. K. (2008). Rural water management in Africa: The impact of customary institutions in Tanzania. Cambria Press.

Nugroho, A. D. (2021). Agricultural market information in developing countries: A literature review. *Agricultural Economics (Zemědělská ekonomika), 67*(11), 468–477. https://doi.org/10.17221/129/2021-agricecon

Obeng-Odoom, F. (2012). Land reforms in Africa: Theory, practice, and outcome. *Habitat International, 36*(1), 161–170. https://doi.org/10.1016/j.habitatint.2011.07.001

Odongo, D. (2023). Four priorities for gender-responsive agricultural policies. [online]. Available at: https://farmingfirst.org/2023/05/four-priorities-for-gender-responsive-agricultural-policies/ Accessed 10 Mar 2024.

Odonkor, E., Hamasi, L., Muthoni, F., Khau, M., & Mutiso, M. (2018). Gender in water finance: Perspectives for a paradigm shift in water finance management. In *Innovative water finance in Africa: A guide for water managers.* Lightning Source.

Ogechi, N. O. (2019). Ethnicity, language, and identity in Kenya. *Modern Africa: Politics, History and Society, 7*(1), 111–135. https://doi.org/10.26806/modafr.v7i1.265

Ogunlela, Y. I., & Mukhtar, A. A. (2009). Gender issues in agriculture and rural development in Nigeria: The role of women. *Journal of Humanities and Social Sciences, 4*(1), 19–30.

Ojediran, F., & Anderson, A. (2020). Women's entrepreneurship in the global south: Empowering and emancipating? *Administrative Sciences, 10*(4), 87.

Oksala, J. (2018). Feminism, capitalism, and ecology. *Hypatia, 33*(2), 216–234. https://doi.org/10.1111/hypa.12395

Olakojo, S. A. (2017). Gender gap in agricultural productivity in Nigeria: A commodity level analysis. *Economics of Agriculture, 64*(2), 415–435.

O'Laughlin, B. (2007). A bigger piece of a very small pie: Intrahousehold resource allocation and poverty reduction in Africa. *Development and Change, 38*(1), 21–44. https://doi.org/10.1111/j.1467-7660.2007.00401.x

Oluwatusin, F. M., & Adesakin, F. M. (2017). Assessment of the adoption of improved agricultural technologies among Cassava Farmers in Ondo State,

Nigeria. *Life Science Journal, 14*(3), 72–79. https://doi.org/10.7537/marsls j140317.12

Omobitan, O., Khanal, A., & Homey, U. (2019). Examining capital constraints and financing sources: How do small farms meet their agricultural spending and expenses? In *Southern Agricultural Economics Association (SAEA) annual meeting*. Tennessee State University, Nashville.

Omojalaibi, J. A., Okudo, A. G., & Shojobi, D. A. (2019). Are women financially excluded from formal financial services? Analysis of some selected local government areas in Lagos State, Nigeria. *Journal of Economic and Social Thought, 6*(1), 16–47.

Onyalo, O. (2019). Women and agriculture in rural Kenya: Role in agricultural production. *International Journal of Humanities, Art and Social Studies, 4*(4), 1–10.

Osabuohien, E. S., Efobi, U. R., Herrmann, R. T., & Gitau, C. M. W. (2019). Female labor outcomes and large-scale agricultural land investments: Macro-micro evidence from Tanzania. *Land Use Policy, 82*, 716–728. https://doi.org/10.1016/j.landusepol.2019.01.005

Oseni, G., Corral, P., Goldstein, M., & Winters, P. (2015). Explaining gender differentials in agricultural production in Nigeria. *Agricultural Economics, 46*(3), 285–310.

Overseas Development Institute. (2016). *Gender, agriculture and water insecurity*. Overseas Development Institute.

Palacios-Lopez, A., Christiaensen, L., & Kilic, T. (2017). How much of the labour in African agriculture is provided by women? *Food Policy, 67*(February), 52–63. https://doi.org/10.1016/j.foodpol.2016.09.017

Paleo, A. (2017). *What is Ecofeminism?* European Institute of the Mediterranean Nature and Sustainability.

Parker, H., Oates, N., Mason, N., Calow, R., Chadza, W., & Ludi, E. (2016). *Gender, agriculture and water insecurity*. Overseas Development Institute.

Parpart, J. L., Connelly, P., & Barriteau, E. (2000). *Theoretical perspectives on gender and development*. International Development Research Centre.

Perrin, C., & Weill, L. (2022). No man, no cry? Gender equality in access to credit and financial stability. *Finance Research Letters, 47*, 102670. https://doi.org/10.1016/j.frl.2021.102670

Peterman, A., Quisumbing, A., Behrman, J., & Nkonya, E. (2011). Understanding the complexities surrounding gender differences in agricultural productivity in Nigeria and Uganda. *Journal of Development Studies, 47*(10), 1482–1509.

Peters, P. (2018). Revisiting the social bedrock of kinship and descent in Africa. In *A companion to the anthropology of Africa* (pp. 33–62).

Phiri, A., & Doku, I. (2024). Is climate finance aiding food security in developing countries? A focus on Sub-Sahara Africa. *Cogent Economics & Finance, 12*(1), 2312777.

Pierotti, R., Clara, D., & Rajdev, K. B. (2022). *Time and money: A study of labor constraints for female cotton producers in Côte d'Ivoire*. https://www.researchgate.net/publication/351980457_Time_and_Money_A_Study_of_Labor_Constraints_for_Female_Cotton_Producers_in_Cote_d%27Ivoire.prevention-collaborative.org

Poulton, C. (2012). Democratisation and the political economy of agricultural policy in Africa. Future Agricultures Working Paper 043. https://eprints.soas.ac.uk/20548/1/FAC_Working_Paper_043.pdf

Presbey, G. M. (2013). Women's empowerment: The insights of Wangari Maathai. In A. Swanson, D. M. Haybron, & J. J. Begley (Eds.), *Happiness and well-being: Integrating research across the disciplines*. University of North Carolina.

Périvier, H. (2023). Gendering the commons or commoning gender studies? A bibliometric analysis. Sciences Po OFCE Working Paper.

Perkins, P. E. (2017). Sustainable commons governance and climate justice: Ecofeminist insights and indigenous traditions. In *Environmental Justice in the Anthropocene-2017*. Colorado State University Libraries.

Petr, B., Libor, G., & Nahanga, V. (2018). Towards a common agricultural policy in Africa? *Agricultural Economics (Zemědělská ekonomika), 64*(7), 301–315. https://doi.org/10.17221/310/2016-agricecon

Phillips, M. E. (2020). Reconnecting with nature: An ecofeminist view of environmental management. *Geographical Research, 58*(2), 154–166. https://doi.org/10.1111/1745-5871.12388

Phiri, A. T., Toure, H. M. A. C., Kipkogei, O., Traore, R., Afokpe, P. M. K., & Lamore, A. A. (2022). A review of gender inclusivity in agriculture and natural resources management under the changing climate in sub-Saharan Africa. *Cogent Social Sciences, 8*(1). https://doi.org/10.1080/23311886.2021.2024674

Porter, G., Hampshire, K., Abane, A., Munthali, A., Robson, E., De Lannoy, A., Tanle, A., Owusu, S. (2019). Mobile phones, gender, and female empowerment in sub-Saharan Africa: Studies with African youth. *Information Technology for Development, 26*(1), 180–193. https://doi.org/10.1080/02681102.2019.1622500

Poulton, C. (2012). *Democratisation and the political economy of agricultural policy in Africa*. Political Economy of Agricultural Policy in Africa. https://opendocs.ids.ac.uk/opendocs/handle/20.500.12413/2245

Poulton, C., & Chinsinga, B. (2018). *The political economy of agriculture commercialisation in Africa*. Agricultural Policy Research in Africa.

Poulton, C., & Kanyinga, K. (2014). The politics of revitalising agriculture in Kenya. *Development Policy Review, 32*(s2), 151–172. https://doi.org/10.1111/dpr.12080

Puleo, A. H. (2017). What is ecofeminism. *Quaderns de la Mediterrània, 25*, 27–34.

Pyburn, R., & Van Eerdewijk, A. (2021). *Advancing gender equality through agricultural and environmental research: Past, present, and future*. Intl Food Policy Res Inst.

Quansah, K. E., Gordon, C., Aboagye-Antwi, F., & Gerchie, R. (2019). The implication of climate variability on household water management; perception and practices among rural womenfolk in four communities in Ghana. *Cogent Environmental Science, 5*(1), 1–16.

Quigley, C. F., Che, S. M., Achieng, S., & Liaram, S. (2017). "Women and the environmental are together": Using Participatory Rural Appraisal to examine gendered tensions about the environment. *Environmental Education Research, 23*(6), 773–796.

Quisumbing, A., Cole, S., Elias, M., Faas, S., Galiè, A., Malapit, H., Meinzen-Dick, R., Myers, E., Seymour, G., & Twyman, J. (2023). Measuring women's empowerment in agriculture: Innovations and evidence. *Global Food Security, 38*. https://doi.org/10.1016/j.gfs.2023.100707

Quisumbing, A., Heckert, J., Faas, S., Ramani, G., Raghunathan, K., Malapit, H., Malapit, H., Heckert, J., Eissler, S., Faas, S., Martinez, E., Myers, E., Pereira, A., Quisumbing, A., Ragasa, C., Raghunathan, K., Rubin, D., & Seymour, G. (2021). Women's empowerment and gender equality in agricultural value chains: Evidence from four countries in Asia and Africa. *Food Security*. https://doi.org/10.1007/s12571-021-01193-5

Quisumbing, A. R., Meinzen-Dick, R. S., & Malapit, H. J. (2019) 'Gender equality: Women's empowerment for rural revitalization', Chap. 5. In *2019 Global Food Policy Report* (pp. 44–51). International Food Policy Research Institute.

Ragsdale, K., Read-Wahidi, M., Méndez, G. R., & Lower, K. (2019). Why gender and landmatter: Examples from rural Ghana. *African Journal of Food, Agriculture, Nutrition and Development, 19*(5), 15155–15158.

Raimi, M. O., Bilewu, O. O., Adio, Z., & Halimat, A. (2019). Women contributions to sustainable environments in Nigeria. *Raimi Morufu Olalekan, Bilewu Olaolu Oyinlola, Adio Zulkarnaini Olalekan, Abdulrahman Halimat (2019) Women Contributions to Sustainable Environments in Nigeria. Journal of Scientific Research in Allied Sciences, 5*(4), 35–51.

Rao, M. (2012). Ecofeminism at the crossroads in India: A review. *Dep, 20*(12), 124–142.

Rao, N., Gazdar, H., Chanchani, D., & Ibrahim, M. (2019) Women's agricultural work and nutrition in South Asia: From pathways to a cross-disciplinary, grounded analytical framework. *Food Policy, 82*, 50–62.

Rao, S. (2016). *Indicators of gendered control over agricultural resources: A guide for agricultural policy and research.*

Recha, W. C. (2018). *Local and regional variations in conditions for agriculture and food security in Kenya.* Egerton Unviersity.

Rierio, A. (2023). The commons. In *Encyclopedia of the social and solidarity economy.* Edward Elgar Publishing.

Rodgers, Y., & Akram-Lodhi, H. (2019). The gender gap in agricultural productivity in sub-Saharan Africa: Causes, costs and solutions.

Rose, D. C., Wheeler, R., Winter, M., Lobley, M., & Chivers, C. -A. (2021). Agriculture 4.0: Making it work for people, production, and the planet. *Land Use Policy, 100*, 104933. https://doi.org/10.1016/j.landusepol.2020.104933

Sabourin, E. (2023). Governance of the commons and social values: A dialogue between Elinor Ostrom's work and the francophone theory of reciprocity. *The Journal of Peasant Studies, 50*(7), 2685–2700. https://doi.org/10.1080/03066150.2022.2120809

Sachs, C. E. (2014). *Women working in the environment: Resourceful natures.* Routledge.

Said, M. K. (2023). Governing the commons: An appraisal of Ostrom principles in the context of community forest management agreement in Zanzibar, Tanzania. *Tanzania Journal of Agricultural Sciences, 3*(9–21), 1–13.

Said, J., & Vencatachellum, V. (2018). *Effective government for Africa's agricultural transformation* (pp. 1–65). Tony Blair Institute for Global Change.

Saka, L., & Adebiyi, O. M. (2021). Women, land dispossession and agricultural production in south-east Nigeria: An eco-feminism perspective. *African Journal of Land Policy and Geospatial Sciences, 4*(3), 423–436.

Salo, E. (2023). Food is an African feminist issue. *Matatu, 54*(1), 17–23.

Samadari, A. M. (2017). *Gender responsive land degradation: Neutrality.* United Nations.

Sarkar, A., Argade, S., & Jeeva, C. J. (2018). Development of an index for assessing farm women's access to resources. *International Journal of Current Microbiology and Applied Sciences, 7*(2), 610–618.

Sasser, J. S. (2016). Population, climate change, and the embodiment of environmental crisis. In *Systemic crises of global climate change* (pp. 57–70). Routledge.

Sene, A. (2018). Land-related tensions and women's empowerment in rural Senegal. *Global Journal of Human Social Science Research, 17*(6), 51–60.

Seyfi, S., Hall, C. M., & Vo-Thanh, T. (2020). The gendered effects of statecraft on women in tourism: Economic sanctions, women's disempowerment and

sustainability? *Journal of Sustainable Tourism, 30*(7), 1–18. https://doi.org/10.1080/09669582.2020.1850749
Sharma, S., & Maheshwari, S. (2016). Use of ICT by farm women: A step towards empowerment. *Indian Research Journal of Extension Education, 15*(3), 60–65.
Sharnappa, P. S. (2018). Is ecofeminism doing favoritism to feminism? *Global Journal of Human-Social Science Research, 18*(2), 3–7.
Shepherd, L. J. (2014). *Gender matters in global politics: A feminist introduction to international relations*. Routledge.
Shiva, V. (1988). *Staying alive: Women, ecology and survival in India*. Zed Press.
Shiva, V., & Mies, M. (2014). *Ecofeminism*. Zed Books Ltd.
Simba, A., Tajeddin, M., Dana, L. -P., & Ribeiro, D. E. (2023). Deconstructing involuntary financial exclusion: A focus on African SMEs. *Small Business Economics*. https://doi.org/10.1007/s11187-023-00767-1
Sinaga, C. A., & Harahap, R. H. (2023). Ecofeminist perspective: Climate change and its impact on women. *Journal of Sumatera Sosiological Indicators, 2*(2), 193–199.
Singh, S. (2007). Deconstructing 'gender and development' for 'identities of women'. *International Journal of Social Welfare, 16*(2), 100–109. https://doi.org/10.1111/j.1468-2397.2006.00454.x
Sinyolo, S. A., Sinyolo, S., Mudhara, M., & Ndinda, C. (2018). Gender differences in water access and household welfare among smallholder irrigators in Msinga Local Municipality, South Africa. *Journal of International Women's Studies, 19*(5), 129–146.
Sitko, N. J., & Jayne, T. S. (2014). Structural transformation or elite land capture? The growth of "emergent" farmers in Zambia. *Food Policy, 48*, 194–202. https://doi.org/10.1016/j.foodpol.2014.05.006
Slavchevska, V., Kaaria, S., & Taivalmaa, S. L. (2016). *Feminization of agriculture in the context of rural transformations: What is the evidence?*. World Bank.
Sokile, C. (2005). Towards improvements of institutional frameworks for intersectoral water management. *The case of Mkoji sub catchment of the Great Ruaha River in the Rufiji Basin, Tanzania*. A thesis submitted in fulfillment of the requirements for the degree of Doctor of Philosophy (Development Studies) of the University of Dar-es-Salaam. University of Dar-es-Salaam.
Sraboni, E., & Quisumbing, A. R. (2018) Women's empowerment in agriculture and dietary quality across five South Asian countries. *Agricultural Economics, 49*(6), 729–744. https://doi.org/10.1111/agec.12461
Srivastava, A. K. (2017). Negotiating femininity with masculinity in water crisis: An ecofeminist perusal of Sarah Joseph's gift in green. *Research Journal of English Language and Literature, 5*(4), 152–155.

Stevano, S. (2021). Classes of working women in Mozambique: An integrated framework to understanding working lives. *Review of International Political Economy*. https://doi.org/10.1080/09692290.2021.1892797

Sudha, T., & Reshi, I. A. (2023). Unleashing the power: Empowering women for a stronger economy. *International Journal of Educational Review, Law And Social Sciences (IJERLAS), 3*(3), 826–833.

Sulaiman, F. (2018). Empowering women farmers through multi-channel learning: An application of need assessment technique. *Indonesian Journal of Agricultural Science, 1*(2), 39–50.

Sulaimanova, B., & Jasoolov, D. (2018). The gender gap in agricultural productivity in Kyrgyzstan. *Uluslararası Avrasya ekonomileri konferansı*. https://doi.org/10.36880/c10.02039

Suresh, L. (2021). Understanding the relationship between sustainability and ecofeminism in an Indian context. *Journal of Developing Societies, 37*(1), 116–135.

Sweetman, C., & Ezpeleta, M. (2017). Introduction: Natural resource justice. *Gender & Development, 25*(3), 353–366. https://doi.org/10.1080/13552074.2017.1395138

Sydee, J., & Beder, S. (2001). Ecofeminism and globalism: A critical appraisal. *Democracy and Nature, 7*(2), 281–302.

Tan, C. (2017). Rebellious women in men's dystopia: Katniss and Furiosa. *Pamukkale University Journal of Social Sciences Institute, 2017*(26), 32–46. https://doi.org/10.5505/pausbed.2017.22599

Tantoh, H. B., McKay, T. T., Donkor, F. E., & Simatele, M. D. (2021). Gender roles, implications for water, land, and food security in a changing climate: A systematic review. *Frontiers in Sustainable Food Systems, 5*, 707835.

Tibesigwa, B., & Visser, M. (2016). 'Assessing gender inequality in food security among small-holder farm households in urban and rural South Africa'. *World Development, 88*, 33–49. https://doi.org/10.1016/j.worlddev.2016.07.008

Tiwari, P., & Malati, N. (2023). Role of training in women empowerment: An empirical analysis. *Journal of Technical Education and Training/Journal of Technical Education and Training, 15*(1). https://doi.org/10.30880/jtet.2023.15.01.020

Travnicek, J., Schlatter, B., & Willer, H. (2024). The world of organic agriculture 2024: Summary. In *The world of organic agriculture*. Statistics and Emerging Trends 2024. Research Institute of Organic Agriculture FiBL, Frick, and IFOAM.

Ts'ehlo, N. J., & Musingati, M. (2024) 'Feminist policies and practices', Chap. 7. In *Advances in psychology, mental health, and wellbeing*. IGI Global.

Udoh, A. (2016). Assessing crop production technologies adoption by women farmers: Some empirical evidence. *Indian Research Journal of Extension Education, 16*(1), 120–123.

Ugwu, P. (2019). *Women in agriculture: Challenges facing women in African farming*. Project report of African women in agriculture.

Umbadda, S., & Elgizouli, I. (2018). Foreign aid and sustainable agriculture in Africa. *Aid Effectiveness for Environmental Sustainability*, 369–395.

Unay-Gailhard, I., & Bojnec, Š. (2021). Gender and the environmental concerns of young farmers: Do young women farmers make a difference on family farms?. *Journal of Rural Studies, 88*, 71–82. https://doi.org/10.1016/j.jrurstud.2021.09.027

UN Women. (2015). *The cost of the gender gap in agricultural productivity in Malawi, Tanzania, and Uganda*. UN Women.

Van Koppen, B., Giordano, M., & Butterworth, J. (Eds.). (2016). *Community-based water law and water resource management reform in developing countries*. Routledge.

Van Koppen, B. (2017a). *Expert paper on rural women's rights to water for health, food and income*. UN Women, IFAD, FAO, WFP Expert Group Meeting 'Challenges and opportunities in achieving gender equality and the empowerment of rural women and girls'. International Water Management Institute.

Van Koppen, B. (2017b). *Rural women's rights to water for health, food and income*. International Water Institute.

Van Koppen, B. (2023). Restoring the commons: A gendered analysis of customary water tenure in Sub-Saharan Africa. *The International Journal of the Commons, 17*(1), 1–11. https://doi.org/10.5334/ijc.1164

Van Rooyen, A. F., Ramshaw, P., Moyo, M., Stirzaker, R., & Bjournlund, H. (2017). Theory and application of agricultural innovation platforms for improved irrigation scheme management in Southern Africa. *International Journal of Water Resources Development, 33*(5), 804–823.

Venkatramanan, V., Shah, S., & Prasad, R. (2019). *Global climate change and environmental policy: Agriculture perspectives*. Springer.

Von Kotze, A., & Walters, S. (2023). Taking risks: Exploring ecofeminist climate just popular education. *European Journal for Research on the Education and Learning of Adults, 14*(1), 17–33.

Waha, K., van Wijk, M. T., Fritz, S., See, L., Thornton, P. K., Wichern, J., & Herrero, M. (2018). Agricultural diversification as an important strategy for achieving food security in Africa. *Global Change Biology, 24*(8), 3390–3400.

Waiganjo, M. M., & Waweru, B. N. (2018). Improving agricultural productivity through effective teaching of agriculture science to girls in secondary schools, Kenya. *International Journal of Innovative Research and Advanced Studies, 5*(11), 24–28.

Waithanji, E. M. (2021). *Kenya's draft agriculture sector Gender Policy: Policy Note*. CGIAR.

Wamboye, E. F. (2023). *The paradox of gender equality and economic outcomes in Sub-Saharan Africa: The role of land rights*. Cambridge University Press.

Wane, N., & Chandler, D. (2002). "African women, cultural knowledge, and environmental education with a focus on Kenya's indigenous women". *Canadian Journal of Environmental Education, 7*(1), 86–98.

Wheeler, S. A., Zuo, A., Bjornlund, H., Mdemu, M. V., van Rooyen, A., & Munguambe, P. (2016). An overview of extension use in irrigated agriculture and case studies in south-eastern Africa. *International Journal of Water Resources Development, 33*(5), 755–769.

Wily, L. A. (2018). 'Risks to the sanctity of community lands in Kenya: A critical assessment of new legislation with reference to forestlands'. *Land Use Policy, 75*, 661–672.

Womin Collective. (2017). Extractives vs development sovereignty: Building living consent rights for African women. *Gender & Development, 25*(3), 421–437.

World Bank. (2015). *The cost of the gender gap in agricultural productivity in Malawi, Tanzania, and Uganda*. World Bank.

World Bank. (2018). *World Development Report 2018: Learning to realize education's promise*. World Bank. Available at: https://openknowledge.worldbank.org/handle/10986/28340

World Bank. (2023). *Gender & agriculture in Sub-Saharan Africa: Review of constraints and effective interventions*. World Bank, Gender Innovation Lab.

World Bank. (2017). *Enabling the business of agriculture*. World Bank.

Wulandari, E., Meuwissen, M. P. M., Karmana, M. H., & Oude Lansink, A. G. J. M. (2017). Access to finance from different finance provider types: Farmer knowledge of the requirements. *PLOS ONE, 12*(9), 179–285.

Yami, M., Abioye, O., Sore, S. Z., Mugisho, A., & Abdoulaye, T. (2024). Factors influencing gender and youth integration in agricultural research and innovation in Africa. *CABI Agriculture and Bioscience, 5*(1), 1–12.

Yee, S., Shin, M. N., Nadia, R., Nurul, A. Y., Munifah S., Abdulmumini, U. (2024). The contribution of agricultural extension to empowerment of women for agricultural development. *Journal of Agricultural Extension, 28*(2), 1–19.

Yeyouomo, A. K., Asongu, S. A., & Agyemang-Mintah, P. (2023). Fintechs and the financial inclusion gender gap in Sub-Saharan African countries. *Women's Studies International Forum, 97*, 102695. https://doi.org/10.1016/j.wsif.2023.102695

You, L., Xie, H., Wood-Sichra, U., Guo, Z., & Wang, L. (2014). Irrigation potential and investment return in Kenya. *Food Policy, 47*, 34–45. https://doi.org/10.1016/j.foodpol.2014.04.006

Yu, Z., & Osabohien, R. (2023). Investigating the impact of education and women in agriculture on human capital outcomes in Africa: An interaction analysis. *Problems of Sustainable Development, 18*(2), 242–248. https://doi.org/10.35784/preko.4037

Zossou, E., Arouna, A., Diagne, A., & Agboh-Noameshie, R. A. (2017). Gender gap inacquisition and practice of agricultural knowledge: Case study of rice farming in west Africa. *Experimental Agriculture, 53*(4), 566–577.

Zwarteveen, M. Z. (1997). Water: From basic need to commodity: A discussion on gender and water rights in the context of irrigation. *World Development, 25*(8), 1335–1349.

Index

M. O. Jimoh and S. Narsiah, *Unveiling the Power of Women in Irrigation*, https://doi.org/10.1007/978-3-031-94174-0

F

G

I

K

L

Kolibri 360 GmbH
Ferdinand-Jühlke-Straße 7
99095 Erfurt, Deutschland
produktsicherheit@kolibri360.de